The Health Care Ethics Consultant

Contemporary Issues in Biomedicine, Ethics, and Society

The Health Care Ethics Consultant, edited by ***Françoise E. Baylis,*** 1994
When Medicine Went Mad: ***Bioethics and the Holocaust,*** edited by ***Arthur L. Caplan,*** 1992
Compelled Compassion: ***Government Intervention in the Treatment of Critically Ill Newborns,*** edited by ***Arthur L. Caplan, Robert H. Blank, and Janna C. Merrick,*** 1992
New Harvest: *Transplanting Body Parts and Reaping the Benefits,* edited by ***C. Don Keyes,*** 1991
Aging and Ethics, edited by ***Nancy S. Jecker,*** 1991
Beyond Baby M: ***Ethical Issues in New Reproductive Techniques,*** edited by ***Dianne M. Bartels, Reinhard Priester, Dorothy E. Vawter, and Arthur L. Caplan,*** 1989
Reproductive Laws for the 1990s, edited by ***Sherrill Cohen and Nadine Taub,*** 1989
The Nature of Clinical Ethics, edited by ***Barry Hoffmaster, Benjamin Freedman, and Gwen Fraser,*** 1988
What Is a Person?, edited by ***Michael F. Goodman,*** 1988
Advocacy in Health Care, edited by ***Joan H. Marks,*** 1986
Which Babies Shall Live?, edited by ***Thomas H. Murray and Arthur L. Caplan,*** 1985
Alzheimer's Dementia: ***Dilemmas in Clinical Research,*** edited by ***Vijaya L. Melnick and Nancy N. Dubler,*** 1985
Feeling Good and Doing Better, edited by ***Thomas H. Murray, Willard Gaylin, and Ruth Macklin,*** 1984
Ethics and Animals, edited by ***Harlan B. Miller and William H. Williams,*** 1983
Profits and Professions, edited by ***Wade L. Robison, Michael S. Pritchard, and Joseph Ellin,*** 1983
Visions of Women, edited by ***Linda A. Bell,*** 1983
Medical Genetics Casebook, by ***Colleen Clements,*** 1982
Who Decides?, edited by ***Nora K. Bell,*** 1982
The Custom-Made Child?, edited by ***Helen B. Holmes, Betty B. Hoskins, and Michael Gross,*** 1980
Birth Control and Controlling Birth: ***Women-Centered Perspectives,*** edited by ***Helen B. Holmes, Betty B. Hoskins, and Michael Gross,*** 1980

The Health Care Ethics Consultant

Edited by

Françoise E. Baylis

University of Tennessee, Knoxville, TN

Humana Press • Totowa, New Jersey

999 Riverview Dr., Suite 208
Totowa, NJ 07512

Printed in the United States of America. 9 8 7 6 5 4 3 2 1

Library of Congress Cataloging in Publication Data

The Health care ethics consultant / edited by Françoise E. Baylis.
p. cm. -- (Contemporary issues in biomedicine, ethics, and society)
Includes bibliographical references and index.
ISBN 0-89603-278-7
1. Medical ethics consultation. I. Baylis, Françoise E., 1961–
II. Series.
R724.H343 1994
610.69'5--dc20 94-13467
CIP

Preface

The primary objective of *The Health Care Ethics Consultant* is to focus attention on an immediate practical problem: the role and responsibilities, the education and training, and the certification and accreditation of health care ethics consultants.

The principal questions addressed in this book include: Who should be considered health care ethics consultants? Whom should they advise? What should be their responsibilities and what kind of training should they have? Should there be some kind of accreditation or certification program to ensure that those who call themselves ethics consultants are in fact qualified to advise, consult, research, and write in health care ethics?

The distinguished authors of these articles are persons with diverse backgrounds, interests, presumptions, and values. Not surprisingly, therefore, diverse responses have emerged to the questions posed.

Though the book's chapters are individually authored, they are informed by the group discussions that went on during active workshop sessions, and by knowledge of the contributions of others. All of the chapters meaningfully represent their consensus. This is not to say that there were no disagreements regarding specific details, but rather that there were no fundamental objections on the book's basic content among a panel of authors who share basic premises regarding the role, responsibilities, education, and certification of health care ethics consultants.

Françoise E. Baylis

Origins of This Book

In the Fall of 1990, a proposal for a Strategic Research Network on "Health Care Ethics Consultation" was submitted to the Social Sciences and Humanities Research Council of Canada (SSHRC).

The grant application was successful and, in April 1991, a national, multi-institutional, and multidisciplinary Network was subsequently established. The Atlantic region, Québec, Ontario, and the West were represented. There were faculty members from Dalhousie University, McGill University, the University of Toronto, McMaster University, and the University of Calgary. As well, there were members with expertise in the areas of philosophy, theology, medicine, and law; and within these disciplines there were individuals with diverse training and professional work experiences. For example, in the area of philosophy there were persons with strong academic backgrounds in applied ethics, but limited practical experience, and there were others with considerable experience in the health care setting and less exposure to academic applied ethics. In the area of theology, there were persons with either "academic" or "professional" degrees. Also, there were important differences in primary roles between the two physician members: one member being first and foremost a health care practitioner, the other being first and foremost a health care ethics consultant. The heterogeneity of the group significantly contributed to the richness of the discussions and the Network's final report.

For two years, members of the Network met to discuss and debate on conflicting viewpoints. The objective was not to achieve consensus for the sake of consensus, but rather to develop a genuine understanding of the different viewpoints and, in the process, to note any points of agreement.

This is particularly worthy of mention because members of the Network participated in sometimes difficult discussions secure in the knowledge that they were free to write a dissent and that all dissenting opinions would be published with any final report. At the end of two years, instead of a series of opinion papers and dissents, there is a collection of essays authored by various people who share many basic premises regarding health care ethics consultants.

The two years of discussion and writing were an invaluable learning experience for members of the Network—an experience that certainly confirmed an initial intuition regarding the value of open, multidisciplinary research. Members of the Network learned a great deal from each other, and what has come out of the project is much more than the written record, which, in itself, is an important testimony to the group's collegial work.

Acknowledgments

Most of the research for this book was funded by grant #806-91-0016, awarded by the Social Sciences and Humanities Research Council (SSHRC) of Canada for a Strategic Research Network on Health Care Ethics Consultation. As well, support for this project was provided by Osler, Hoskin & Harcourt through their Community Law Program, and by the Research Institute at The Hospital for Sick Children through their Summer Student Program.

Particular thanks are owed to Micheline Cox for her administrative assistance in coordinating meetings and mailings for the Strategic Research Network. Special thanks are also offered to Michael Bolton, Claudia D'Souza, Anna Glasgow, T. Douglas Kinsella, Louise Kunicki, Edna McHutchion, Ariella Pahlke, Sharon Rae, Jennifer M. Smith, Lynda Sullivan, and Charles E. Weijer for their varied contributions to this project.

Contents

Contributors

FRANÇOISE E. BAYLIS • *Department of Philosophy, University of Tennessee, Knoxville, TN*

EUGENE BEREZA • *McGill Centre for Medicine, Ethics, and Law, Montreal, Canada*

MICHAEL BURGESS • *Department of Medical Ethics, University of Calgary, Canada*

BRIDGET CAMPION • *PhD Candidate, Regis College, Toronto School of Neology, Toronto, Canada*

MICHAEL D. COUGHLIN • *St. Joseph's Hospital, Hamilton, Canada*

JEANNE DESBRISAY • *Osler, Hoskin & Harcourt, Toronto, Canada*

JOCELYN DOWNIE • *Law Clerk, Supreme Court of Canada, 1993–1994, Ottawa, Canada*

BENJAMIN FREEDMAN • *McGill Centre for Medicine, Ethics, and Law, Montreal, Canada*

LARRY LOWENSTEIN • *Osler, Hoskin & Harcourt, Toronto, Canada*

ABBYANN LYNCH • *Bioethics Department, Hospital for Sick Children, Toronto, Canada*

SUSAN SHERWIN • *Department of Philosophy, Dalhousie University, Halifax, Nova Scotia*

JANET STORCH • *Faculty of Nursing, University of Calgary, Canada*

JOHN L. WATTS • *McMaster University Medical Centre, Hamilton, Canada*

GEORGE WEBSTER • *Department of Clinical Ethics, St. Michael's Hospital, St. Joseph's Health Centre, and Providence Centre, Toronto, Canada*

Introduction

Françoise E. Baylis

During the 1970s, those who functioned as bioethicists[1] in the clinical setting typically were either moral philosophers and moral theologians who were invited to apply their knowledge and skills to the health care setting, or physicians whose medical colleagues believed them to be particularly skilled at resolving "hard cases." Initially, the contribution of these bioethicists was limited to sporadic case consultation and occasional committee work at a supra-institutional level.

For the "nonmedically-trained bioethicists," case consultation was an occasional act, incidental to their primary occupation. For the "medically trained bioethicists," case consultation was perhaps more frequent, but still sporadic. In a relatively short period of time, however, requests for case consultation became more frequent, along with requests for ethics committee work (clinical or research), ethics education, and ethics input into policy formulation. As these responsibilities accumulated, the position of bioethicist was formalized.[2]

The development of "career lines for would-be bioethicists,"[3] the introduction of speciality journals, books, courses, and conferences, the founding of the Hastings Center, and the increasing number of invitations to participate in public policy debates signaled the emergence of a new professional practice. This, in turn, suggested a need for specialized training programs in bioethics. Such programs for graduate and postgraduate students in philosophy, theology, and medicine were developed in the mid to late 1970s by individuals with informal postcareer training in bioethics and others with an academic interest in applied ethics. Generally, the programs were designed in such a way that those with academic training in ethics and moral reasoning were to be provided with clinical experience, whereas those with clinical experience were to be provided with knowledge of ethics and skills in moral reasoning.[4] Furthermore, both groups of trainees were to be provided with other relevant educational experiences, as appropriate (e.g., health law). The different programs, however, essentially preserved the character and curriculum of their home discipline, and students were expected to meet the established standards of their respective disciplines. With time, graduate specialized training in bioethics increasingly became the accepted means of entry to the practice for those with a background in philosophy, theology, or medicine. Meanwhile, individuals in other disciplines (e.g., law, physiotherapy, nursing, religious studies, and sociology) continued to join the ranks on the basis of informal postcareer training.

For some years now, graduates of specialized programs in bioethics and medical ethics have worked in various health care settings alongside those with informal postcareer training. These two groups of health care ethics consultants, however, differ in significant respects. For example, health care ethics consultants with informal postcareer training are persons with a prior career (and often established credibility) in a related profession. By comparison, health care ethics consultants with specialized training typically have no prior career in a related discipline (and hence, no prior established credibility). Unfortunately, the absence of prior established credibility, the comparative lack of clinical experience (especially for those who are not health care practitioners), and differences in age and gender[5] have all contributed to a difference in authority and effectiveness for health care ethics consultants with specialized training, as compared with health care ethics consultants with informal postcareer training.

Presently, as those with specialized training struggle to establish their credibility both in their home discipline and in ethics consultation, a new breed of health care ethics consultant has begun to join the ranks. The difference in background and training between this group of ethics consultants and the previous two is considerable. First, there is no necessary "home discipline," but instead there is a variety of "means of entry" to the practice. Second, there is no certification or credentialing at the doctoral or fellowship level in a related discipline accompanied by specialized training, but instead a wide range of available experiences including: self-directed learning and conference attendance; one-week intensive bioethics courses;[6] 13-week courses (two hours a week) with a limited apprenticeship;[7] and one- and two-year M.A. specializations in medical ethics.[8] Third, there is no core of professional activities, but rather a design on a career path that involves perhaps only case consultation, only committee work, only ethics education, or only policy formulation.

The widening difference in background, training, and practice among those who promote themselves as health care ethics consultants brings into sharp focus a point made recently by Jonathan Moreno—namely, that health care ethics consultation, like other nascent professional practices, "is a professional pursuit that has grown in demand without a clear prior understanding of precisely what qualifies one to provide this service."[9]

As a wide variety of people with an increasingly diverse set of qualifications present themselves as health care ethics consultants, demands are being heard for the establishment of professional standards governing entry into this field.[10] Many ethics consultants join with various (prospective) clients and employers in the desire to establish standards and uniform measures of professional qualification. This is not surprising since within the health care setting, certification of professional status is the norm, so it is a familiar and attractive model.

In the chapter by Susan Sherwin, the author explores the appropriateness of developing a formal process of certification or accreditation for individuals who are deemed qualified to practice as health care ethics consultants. She argues that the certification of health care ethics consultants is at best premature and at worst self-serving. There are clear dangers in moving too quickly to a formal certification or accreditation process designed to screen out those whose training or experience does not conform to some preestablished norm. Ethics is a different sort of subject matter from that at the core of other health-related specialities, and it is by no means obvious that expertise in ethics consultation can be adequately captured by the same sorts of processes that work in other health professions. Sherwin suggests that other models and strategies for ensuring "quality control" of health care ethics consultants may serve as alternatives to the more familiar route of professional certification.

Next, Françoise Baylis provides a functional description of the health care ethics consultant. This Profile of a Health Care Ethics Consultant (Profile) outlines the requisite knowledge, abilities, and traits of character for case consultation on ethical issues in clinical care or in clinical research, and for ethics consultation to clinical ethics committees, research ethics boards (institutional review boards), and policy formulation committees in health care institutions.

The Profile of the health care ethics consultant is proposed as an alternative to certification. It is provided for use by students, educators, prospective employers, and others. Of necessity, a certification mechanism functions from the top down. This potentially raises questions about who controls access to professional practice. An advantage of the Profile as an alternative to certification is that it can function bottom up as well as top down. For example, a student can use the Profile to assess an educational program: Does the program provide training that will enable me to meet the standards of practice outlined in the Profile? An educational institution can use the Profile for self-assessment: Does our existing program provide graduate students with an opportunity to acquire/develop the requisite knowledge, abilities, and traits of character?[11]

Also, a health care administrator (prospective employer) can use the Profile to assess job candidates for a position as a health care ethics consultant, or to determine appropriate membership for the health care ethics committee. A careful consideration of the institution's needs will allow the administrator (prospective employer) to identify those aspects of the Profile relevant for practice within the institution. On this basis, the administrator (prospective employer) can ask one or both of the following questions: Does the ethics committee as a whole (or the ethics consultant) have the qualifications and resources to meet the institution's needs? Does this particular job candidate have the qualities needed for ethics consultation in this institution? This being said, the Profile of the health care ethics consultant is not a job description, but rather an inventory of the requisite knowledge, abilities, and traits of character for effective health care ethics consultation.

Among the knowledge and abilities listed in the Profile are: extensive knowledge of facilitation techniques including mediation, negotiation, and arbitration; and the ability to use one or more of these techniques as appropriate.

In the chapter by Abbyann Lynch, the author explores in some detail these two elements of the Profile, underlining the importance of mediation as an effective means of consultation on ethical issues in clinical care. Lynch notes that much of the literature's discussion on health care ethics consultation focuses narrowly on the *substance* of consultation. Consider, for example, the focus on ethical dilemmas arising in the provision of health care or the focus on ethical concerns regarding policy formulation in the health care context. Consistent with such focus, considerable attention is given to the ethics consultant's capability *qua* ethicist. The concern is with theoretical knowledge of ethics, ethical reflection, and ethical problem-solving.

By comparison, concern regarding the health care ethics consultant's *practical* capability is meager, and yet the ethics consultant's practical capability is a necessary complement to his/her substantive capability. Lacking knowledge and skill in the matter of interpersonal communication or group dynamics, the consultant's substantive capability may come to naught. Lynch maintains that due consideration must be given to both aspects of consultation; otherwise, the claim that ethics consultation actually involves consultation is unwarranted.

Still, the questions remain: How are individuals to acquire the requisite knowledge, abilities, and traits of character for effective health care ethics consultation? What training should they have?

Would-be health care ethics consultants have any number of educational experiences available to them, ranging from accredited graduate programs to intensive weekend workshops. Of these, all will introduce students to basic concepts in health care ethics, some will take students into the clinical setting as observers, and others will offer a system of internship. What these programs share is an approach that focuses on a particular task, in this case, producing students with expertise in bioethics. Like certification, it is an approach that insists on fragmentation and specialization.

Just as the Profile stands as an alternative to certification, the feeder disciplines approach, outlined in the chapter coauthored by Michael Burgess, Eugene Bereza, Bridget Campion, Jocelyn Downie, Janet Storch, and George Webster, stands as an alternative to specialized education. Traditionally, health care ethics consultants have come to the field with degrees in such diverse disciplines as law, medicine, nursing, philosophy, and theology. During the course of their studies, they have laid the foundations for their future practice, utilizing the resources of their particular faculties as well as those of the larger university to acquire requisite knowledge and abilities, while meeting their degree requirements. As practitioners, they have come to rely on ongoing multidisciplinary collaboration with colleagues to supplement and sharpen their skills and knowledge.

The chapter on feeder disciplines offers an overview of this approach, written by persons who themselves are products of feeder disciplines and experiences in the field of health care ethics consultation. The guiding belief is that the current tradition is worth preserving.

Describing the qualifications and training needed by health care ethics consultants is, however, at best, half the job. The chapter by Benjamin Freedman tackles the other half: What are the qualifications and conditions necessary for an institution hiring a consultant? What are a consultant's reasonable expectations with respect to job conditions? Contrary to "ethical exceptionalism"—the view that something about the practice of ethics consultation makes it uniquely exempt from the need to specify working conditions—Freedman argues that decent persons working at ethics consultation ("neither slaves nor heros," in his phrase) are entitled to advance arrangements outlining their conditions of employment.

Freedman looks for inspiration to Lon Fuller's work, in trying to outline an "internal morality of bioethics": the minimal conditions necessary for health care ethics consultation to be effective. In the currently unsettled state of definition of the tasks required of ethics consultants, he proposes an interim "thin theory" of working conditions, respecting three tasks that, he argues, will necessarily be essential to a consultant's work: communication; uncertainty of task and authority; and moral integrity. Freedman discusses these tasks in practice and comments on the protections a health care ethics consultant might reasonably require.

The next chapter, coauthored by Larry Lowenstein and Jeanne DesBrisay, considers the potential for legal liability of health care ethics consultants. It does not represent legal advice, but rather the opinions of the authors about the general question whether health care ethics consultants might be found legally liable for the consequences of their consultations. Lowenstein and DesBrisay explore two possible sources of liability—battery and negligence. Following a discussion of the basic elements of battery and negligence and an application of these elements to health care ethics consultation, they conclude that the chances of a court finding a health care ethics consultant liable for damages are fairly small.

The final chapter of the book provides the reader with recent data on the practice of health care ethics consultation in Canada. There is information about who is actually doing ethics consultation, what kind of training and skills they possess, what ethics activities they are engaged in, and what their views are on certification.

Michael Coughlin and John Watts, who analyzed the survey results, found that health care ethics consultants are a very heterogenous group. There is a broad range of formal training and of time commitment to clinical ethics work, a broad range of expertise and wide variation in practices. The results of the survey support the concept of feeder disciplines and the overall preference of the SSHRC Strategic Research Network on Health Care Ethics Consultation (Network) for identifying an ideal Profile, rather than defining a process for certification or accreditation of training.

As a final remark, the Network's reflection on the role and responsibilities, the education and training, and the certification of health care ethics consultants is timely both on the grounds of clarification for those who practice as health care ethics consultants and protection of the public. A concern for practicing ethics consultants, and for those who rely on the services provided by them, is that an increased demand for "ethics assistance" will create lucrative employment opportunities for those with little more than a self-professed interest in health care ethics. To protect "legitimate practitioners" and the public from charlatans, some have suggested a move toward a rigid licensing system. This book is an attempt at a creative alternative, and we hope that the content will resonate well with health care ethics consultants.

Notes and References

[1]In this chapter, "bioethicist" is the historical term that refers to those who originally "did ethics" in the clinical setting; "health care ethics consultant" is the contemporary term that refers to those who currently provide ethics case consultation (clinical or research), ethics committee consultation (clinical or research), ethics education, and policy formulation in the clinical setting. Although descriptively accurate, the term "health care ethics consultant" is not widely used. More common are such terms as bioethicist, ethics consultant, clinical ethicist, medical ethicist, nurse ethicist, and philosopher ethicist. These other descriptive terms are problematic, however, because the focus is either too narrow (e.g., medical ethicist) or too broad (e.g., bioethicist). The term "health care ethics consultant" avoids this problem. For ease of reading, "health care ethics consultant" on occasion is shortened to "ethics consultant."

[2]Note, this version of the history of health care ethics consultation differs from that told by John LaPuma and E. Rush Priest, who summarily note that "ethics consultation emerged as part of clinical ethics, a field of expertise in medicine." John La Puma and E. Rush Priest, "Medical Staff Privileges for Ethics Consultants: An Institutional Model," *Quality Review Bulletin* 18.1 (1992): 17.

[3]David J. Rothman, *Strangers at the Bedside* (New York: Basic, 1991) 189.

[4]E.g., a one-year Fellowship at the University of Chicago, United States originally for physicians and now available to others interested in health care ethics consultation; a minimum four-year doctoral program in philosophy with a formal clinical component at the University of Tennessee, United States (and previously at the University of Western Ontario, Canada).

[5]Health care ethics consultants with informal postcareer training are typically older and, as a group, more predominantly male than health care ethics consultants with specialized training (cf Michael D. Coughlin and John L. Watts, in this vol.).

[6]E.g., Georgetown University, United States.

[7]E.g., University of Virginia, United States.

[8]E.g., McGill University, Canada, and the University of Toronto, Canada.

[9]Jonathan D. Moreno, "Call Me Doctor? Confessions of a Hospital Philosopher," *Journal of Medical Humanities* 12.4 (1991): 194.

[10]Joel Frader and Robert Arnold, "Standards for Clinical Ethics Consultation," *Newsletter of the Society for Bioethics Consultation* Winter (1993): 2,3.

[11]The chapter on "feeder disciplines" begins this application for some of the major contributing academic and professional disciplines.

Certification of Health Care Ethics Consultants

Advantages and Disadvantages

Susan Sherwin

The role of the health care ethics consultant is becoming an increasingly common one in health care practice. For example, many health care institutions and practitioners have found reason to seek either regular or occasional advice from health care ethics consultants. Some health care institutions and policy makers have gone one step further and created positions for ethics consultants, whereas others are considering such a move. In addition, various individuals with a background in health care ethics have expressed interest in offering their services as health care ethics consultants. In most areas of specialized expertise in which qualified individuals offer their services as consultants, aspiring practitioners must establish their credibility and competence by satisfying formally specified programs of professional training and licensure. However, no specific qualifications, standards, or criteria have as yet been established to govern the practice of consultation in the area of health care ethics.

It is no surprise, then, that there is growing interest in the idea of establishing mechanisms and measures to determine who is qualified to offer such consulting services. In the literature on ethics consulting, various authors have proposed that a formal process of certification or accreditation be instituted to set minimum standards of competency, identify qualified specialists, and provide some degree of "quality control" over those who are recognized as qualified to offer their services as health care ethics consultants.[1]

Before endorsing such an initiative, it is important to reflect on the advantages and disadvantages that are associated with any move toward certification or credentialing. This task is complicated by the fact that some of the advantages and disadvantages will often be deeply intertwined—i.e., many of the features that constitute advantages carry with them associated dangers.

To begin with the most obvious advantage, certification offers a uniform means of establishing and maintaining standards of competency for practitioners, and this objective is surely the principal rationale behind proposals to institute the various procedures and criteria that are customarily associated with certification. Certification of qualified individuals would demonstrate their mastery of a specific body of relevant knowledge, and would indicate their attainment of an appropriate level of specialized skills and abilities.

Because certification (or licensing) is the usual means by which other professions establish and maintain standards of specialized expertise and skill, certification of health care ethics consultants would help them to establish their collective status as members of a "profession." This would provide the attendant social status and authority that usually accompanies professionalization, and it would also make explicit the associated responsibilities of professionalization in the form of accountability and self-regulation. Certification may even encourage more use of health care ethics consultants, and it might create more jobs for consultants in hospitals and other institutions; it might also make the field more attractive to talented students who will be encouraged to participate in a more clearly established practice. Also, formal screening and certification procedures would help to establish and protect the reputations of qualified practitioners by providing a mechanism by which they can dissociate themselves from unqualified competitors.

Another advantage of the certification of qualified health care ethics consultants is that it would help employers and clients to identify suitable candidates to meet their particular needs in the area of health care ethics consulting. Moreover, certification would help those who employ consultants to understand what appropriate expectations are when they engage a health care ethics consultant, and it would offer both the consultant and the employer guidance in drawing up job descriptions and contracts. Thus, clarification could reduce much of the tension that is now implicit in such relationships, since setting standards of practice for the profession would make it easier for individual health care ethics consultants to clarify the limits of their expertise. This might thereby reduce the frustration that follows if their employers and clients hold unrealistic expectations.

Further, it is important to acknowledge that health care ethics consultants work in the highly professionalized world of health care delivery. In this environment, certification will help to establish health care ethics consultants as constituting a profession, along a model that is familiar to their colleagues from other professions. It will also be helpful in establishing the credibility of consultants in the minds of other health professionals by setting standards that can be seen as external to individual practitioners. Furthermore, the very process of moving to certification will engage those who are active in health care ethics consulting in an attempt to build consensus on the norms of practice; such a process will in turn provoke healthy debate about what body of knowledge and what abilities are essential to someone practicing under the title of health care ethics consultant. Certification may also facilitate the development of professional organizations that can help monitor and regulate practice, disseminate information, investigate matters of common interest, and so forth. More generally, certification could provide the infrastructure and channels of communication for peer support among practitioners.

In addition, certification and the associated creation of "health care ethics consultant" as a professional designation would, in all likelihood, help to clarify questions of legal liability and the need for obtaining insurance for services offered. It can be anticipated that certification would facilitate the establishment of standards of practice, and in particular, it would encourage the determination of ethical standards governing its practitioners. This is an urgent matter for a newly emerging practice, at a time when health care ethics consultants are faced with the task of achieving recognition and acceptance from other health professions—circumstances that increase the risk of compromising their own moral responsibilities. Trained to identify ethical problems in the delivery of health care, ethics consultants may sometimes find themselves confronted with the choice of either raising ethical alarms, and thereby alienating the very health care practitioners whose confidence they have been trying to win, or remaining silent, and securing their place and authority within the health care system. Silence in the face of moral wrongs is a serious breach of ethics for any profession, but it seems especially objectionable on the part of people whose profession is centerd on questions of ethics. Yet, health care ethics consultants are human, and it is not hard to envision scenarios where long-term acceptance seems more morally (as well as personally) important than immediate protest. Such dangers are present for health care ethics consultants whether or not they are certified; however, certification at least offers a mechanism for setting standards of practice about such matters, and making practitioners accountable not only to their employers and clients, but also to their profession.

Thus, certification certainly seems to be an attractive option for those who are likely to qualify for this status. Moreover, it promises to simplify the quest of patients, physicians, and health care institutions who perceive the need for such services, but need direction in finding reliable consultants (and away from disreputable ones).

Nonetheless, despite its clear attractions, the prospect of certification of health care ethics consultants raises some serious concerns. Instituting common criteria by which standing is achieved in a certified profession carries with it inherent problems. For example, it must be understood that the practice of certification is intentionally exclusionary: Its very purpose in certifying some persons as authorized or legitimate practitioners in the field is to reject all others who might otherwise seek to offer themselves as health care ethics consultants. As with any profession, the effect of excluding some potential practitioners is a restrictive narrowing of the perspectives and practices that will be deemed proper and protected. Certification or licensing within a profession limits the sorts of approaches that will be deemed legitimate in the delivery of services associated with it, and there is always room to dispute the limits that are actually drawn. For example, in the domain of medicine, licensing of physicians in this century has been reserved for practitioners of allopathic medicine, and this decision has undoubtedly stunted the development of homeopathic medicine and other healing modalities; some of these paths not taken may have offered better health care to certain sorts of patients. Loss of nonapproved services—services that may well be beneficial in at least some cases—is one of the costs of certification practices.

In the specific case of health care ethics consultants, certification clearly poses the threat that the line demarcating what is recognized as acceptable practice will be drawn inappropriately; as a result, some individuals who might be very effective in the role of health care ethics consultant will be barred from practice, because they lack the necessary credentials or because their particular orientation is out of fashion at the time. For example, there are currently several competing schools of thought in health care ethics. Some practitioners endorse an approach to problems based on a particular ethical theory (e.g., utilitarianism for Peter A. Singer,[2] and a Kantian style deontology for H. Tristram Englehardt, Jr.[3]); many seem to favor an approach centered on a few central ethical principles (e.g., Tom L. Beauchamp and James F. Childress[4]); others argue for a situationally determined casuist approach (e.g., Albert R. Jonsen and Stephen Toulmin[5]); still others, including myself,[6] argue that ethical analysis cannot be separated from an investigation into broad social and political questions. Further, there are ongoing disputes among practicing ethics consultants about who is qualified to do this sort of work: Some argue that only physicians can be considered appropriate (e.g., John La Puma, Mark Siegler[7]), whereas others suggest that training in philosophy is preferable, if not ideal (e.g., Robert M. Veatch, Singer[8]).

The process of institutionalizing certain approaches as legitimate and declaring others as illegitimate threatens to foster a certain homogeneity in perspective among practitioners that could conflict with the very function of health care ethics consultants by reducing tolerance for certain lines of moral investigation. (For example, it could become entrenched that all consultants are expected to subscribe to a rights-based approach to health care ethics when in fact there are strong ethical arguments against such perspectives.)

In a field like health care ethics, where the "right" approach and the "right" answers are, by their very nature, heavily contested, there are significant grounds for concern that the models available for establishing professional standards of certification are inappropriate to this particular sort of activity. Whereas the adequacy of the standards set for other professions can—at least in theory—be evaluated by objective, empirical data, this is not possible with health care ethics consultation, which typically involves its practitioners in many diverse and largely intangible tasks, such as facilitating multiparty decision making. These tasks do not lend themselves to readily quantifiable measures; thus, it is unclear how "success" in the practice of ethics consultation is to be evaluated. Moreover, whatever constraints govern the practice of certified "experts," they should be subject to ongoing review and criticism. There is, however, no obvious basis on which to resolve many of the difficult ethical disputes that exist among practitioners; there is not even any agreed on methodology available for discussing such issues.

It is also difficult to determine who is qualified to control certified entry to the ranks of health care ethics consultants. Given the open nature of the questions about ethics that lie at its roots, it is desirable to encourage richness and diversity in the approaches taken to health care ethics consulting. At present, there are only a small number of self-selected people engaged in the work of health care ethics consultants. It is far from clear that these particular individuals should be collectively granted the authority to determine who might join their ranks in the future (which is the usual approach to certification). However, it is equally difficult to determine what other system of selection should be chosen.

Professionalization seems, inevitably, to lead to increased bureaucratization and the pursuit of legalistic approaches to problems. Already, some authors are concerned that these trends are evident in health care ethics consulting and may be working to harm patients.[9] In addition, the notion of certification normally carries with it the belief that specific professional training uniquely qualifies its members for certain sorts of tasks. The range of these tasks is still being determined for health care ethics consultants, but some are already clear; one of the principal tasks is that of facilitating decision making on difficult ethical dilemmas that arise in the delivery of health care. So, we must ask whether there is reason to believe that any specific form of training does ensure effectiveness in this role. We should not leap to the conclusion that designated experts are always the best people to turn to for help in resolving problems; many sorts of problems are better handled by the individuals who are closest to them than by appeal to experts. For example, Alcoholics Anonymous (AA) claims success in helping some people overcome their dependence on alcohol through nonprofessional self-help support—often in cases where professional intervention has failed. At least some of the time, it seems that ethical decision making is also better handled through self-help models than through the expert-consultant model.

Moreover, we can anticipate that any formal process of certification that requires health care ethics consultants to possess certain specialized expertise may well instill in its members a sense of proprietary control over the practices in which they are deemed expert. After all, it is common for professionals to promote their own authority and establish their role as indispensable by mystifying their particular knowledge and skills, and by fostering the dependence of others on their services. They often become self protective and preoccupied with pursuing their collective advancement. In their efforts to establish their distinct authority, health care ethics consultants may be led to exaggerate their level of expertise and skill, and suggest greater authority than they really possess.[10] Further, they may become caught up in the familiar interprofessional battles of turf protection common to the health care bureaucracy and find themselves driven to protecting their own primary authority over the areas in which they claim specialized authority. As a result, they may reject the input of colleagues with different training. Such behavior would run counter to the current value placed on diversity in perspective and training that is now widely accepted as desirable in health care ethics.

Indeed, this appreciation of diversity among health care ethics practitioners constitutes an important sign of hope that this specialty will not evolve on the same model as most other established professions, and so it may not be subject to some of the concerns that attach to professions in general. To date, health care ethics consultants have emerged from a variety of different academic and professional backgrounds, and there is wide agreement among many engaged in health care ethics that the work they do requires a significant degree of multidisciplinary training. Health care ethics consulting is, inevitably, an interdisciplinary field. Moreover, its effective practice requires cooperative interactions with a variety of people from a range of different experiences and training. Also, at present, few health care ethics consultants work solely as consultants. Most engage in this activity as an adjunct to their primary role in health care ethics teaching, research, or participation on an ethics committee. Thus, interdisciplinary interactions are an integral part of their training and work. Such experience may limit their enthusiasm for competing with others involved in health care for their own unique turf by rigidly dividing up the services associated with patient care among different specialties.

Nevertheless, when declaring that there are "experts" in health care ethics, we imply that ethical decision making requires a form of technical expertise like surgery, when perhaps it is better to encourage people to approach ethical dilemmas as ordinary, not extraordinary, aspects of their lives. If health care ethics consultants are defined as possessors of some esoteric specialized knowledge, the message conveyed is that the problems they investigate are too difficult for "ordinary" folks, and should regularly be referred to experts who can turn their hard-earned training to the task. People associated with health care delivery may become less willing to assume individual responsibility for the ethical dimensions of their acts and, therefore, less inclined to accept the anguish of moral uncertainty if they perceive ethics as someone else's job. However, health care ethics consultants cannot relieve individuals of their personal responsibility for their own actions, and they must be careful not to give the impression that they do any such thing. Ethics consultants must make it very clear that their job is to help inform and support rather than replace the value decisions of those engaged in providing and receiving health care services.

Finally, it is important to recognize that the debate about certification of health care ethics consultants is going on in a specific socio-economic, political context in which professionalization is deeply connected with questions of class and status. Professionals are all found in the middle class or above, and they and their representative organizations work hard to ensure that, as professionals, they maintain or improve that status. Thus, the perspectives and values reflected by members of the professions are overwhelmingly those of the middle class. All professions are made up of a relatively homogeneous group of well-educated, largely white, Western, able-bodied people with the outlooks that their positions afford them. Establishing health care ethics consulting as a profession will help its practitioners protect their middle-class status, and it may ensure that those who follow will come from similar backgrounds. It is important, then, to remember that theirs are not the only perspectives and values of society; they have a responsibility to seek out information about the perspectives they are habitually insulated from. It surely cannot be acceptable for health care ethics to allow itself to become an instrument for maintaining the hegemony and privilege of its own practitioners; thus, those who call themselves health care ethics consultants will have to work hard to counter the inertia carrying them in such directions.

Despite the undeniable attractions of certification, then, the problems associated with the certification of health care ethics consultants are serious enough to persuade members of the Network that other, more innovative initiatives are preferable means of responding to the pressures for certification. We (the Network members) endorse a less formal, more open-ended approach to establishing standards of training and practice. Specifically, we propose consideration of the "Profile of the Health Care Ethics Consultant"[11] as offering clarification of the knowledge, skills, and character traits that are valuable in those who offer their services as health care ethics consultants, without establishing a single, rigid norm of training for all practitioners. We think that this description of abilities and expertise provides adequate guidance to those who seek the services of someone qualified to serve as an ethics consultant in the area of health care, and it offers direction for study and training to those who seek to acquire the qualifications necessary for competent practice. At the same time, it avoids the rigidity involved in setting one group up as gatekeepers who control those seeking status as legitimate practitioners. Further, it maintains our commitment to different paths as appropriate background for health care ethics consultants, thereby encouraging diversity of views and perspectives among practitioners and ensuring continuing debate about appropriate practices and responses.

We fear, however, that it may already be too late to protect against such a shift in acceptance of moral responsibility. Health care ethicists, if not yet health care ethics consultants, are now a familiar part of the health care landscape and few health care providers remain oblivious of their existence. Moreover, we believe that many of the roles that health care ethicists are asked to play, including that of ethics consultant, are legitimate and valuable. We realize how important it is to both practitioners and employers that appropriate and inappropriate expectations are clearly delimited. Further, we perceive that the demand for certification on the part of both practitioners and clients is growing and that, in practice, it may become irresistible. If those who are active in the profession do not take the initiative to set standards of practice and criteria for qualification, there is a danger that others outside the practice may do it for us in ways that would be even more unsatisfactory. If this profession is to be defined and regulated, we believe it is better that it be self-defined, and not constructed by those who have little or no experience of health care ethics consulting.

Hence, although most of us believe that the disadvantages of certification outweigh the advantages, and would prefer to seek some alternative ways of meeting the legitimate demands represented by a desire for certification, we recognize that circumstances and the desires of others may soon make certification inevitable. If, despite our concerns, certification of health care ethics consultants is ultimately adopted, we hope that the ways in which this "profession" differs from the patterns that are common to others remain clear. In particular, the historical fact that there are many different routes by which people arrive in the role of health care ethics consultant should remain prominent; no single educational program can capture the range of training, experience, and interests now reflected among practitioners. Communication skills, character traits, and intuitive understandings may be as important to effective practice in health care ethics consulting as any particular knowledge base. Since there is no agreed-upon "canon" or specific body of knowledge recognized as essential to effectiveness in the role of health care ethics consultant, it would be improper to institutionalize any specific approach to ethics or bioethics as "fundamental."

Thus, we suggest that any process of certification for prospective practitioners or any system of accreditation for programs that purport to train health care ethics consultants be evaluated according to the criteria we have proposed in the Profile of the health care ethics consultant. We believe that the Profile offers a reasonable standard by which certification or accreditation criteria can be evaluated. It can serve as a meta-standard by which specific evaluative programs can themselves be measured.

Finally, a reminder of the limits inherent in the Profile itself: The Profile is offered as a standard of talent and achievement for those who seek to work as health care ethics consultants. There is a great deal of other activity within the realm of health care ethics beyond consulting. Although there seem to be increasingly strong pressures to find ways of certifying the competence of practitioners of health care ethics consulting, there is far less urgency to certify people in the same way for other activities in the domain of health care ethics. The Profile is designed to cover only the specialization within health care ethics of those who seek to serve as consultants. It is not a reasonable measure for other activities or for programs that offer training in health care ethics generally.

References

[1]*See*, e.g., Joy D. Skeel and Donnie J. Self, "An Analysis of Ethics Consultation in the Clinical Setting," *Theoretical Medicine* 10 (1989): 289–299; John La Puma and E. Rush Priest, "Medical Staff Privileges for Ethics Consultants: An Institutional Model," *Quality Review Bulletin* 18.1 (1992): 17–20; David C. Thomasma, "Why Philosophers Should Offer Ethics Consultations," *Theoretical Medicine* 12 (1991): 129–140; Sigrid Fry-Revere, *The Accountability of Bioethics Committees and Consultants* (Frederick, MD: University Publishing Group, 1992).

[2]Peter A. Singer, *Practical Ethics* (Cambridge: Cambridge University Press, 1979).

[3]H. Tristram Englehardt, Jr., *The Foundations of Bioethics* (New York: Oxford University Press, 1986).

[4]Tom L. Beauchamp and James F. Childress, *Principles of Biomedical Ethics*, 3rd ed. (New York: Oxford University Press, 1989).

[5]Albert R. Jonsen and Stephen Toulmin, *The Abuse of Casuistry: A History of Moral Reasoning* (Berkeley: University of California Press, 1988).

[6]Susan Sherwin, *No Longer Patient: Feminist Ethics and Health Care* (Philadelphia: Templeton University Press, 1992).

[7]John La Puma and David L. Schiedermayer, "Ethics Consultation: Skills, Roles, and Training," *Annals of Internal Medicine* 114.2 (1991): 155–60; Mark Siegler, "Clinical Ethics and Clinical Medicine," *Archives of Internal Medicine* 139 (1979): 915.

[8]Robert M. Veatch, *A Theory of Medical Ethics* (New York: Basic, 1981) 8; Peter A. Singer, "Moral Experts," *Analysis* 32.4 (1972): 115–117.

[9]*See* Colleen D. Clements, "Was Socrates a Philosophy-Basher," *APA Newsletter on Philosophy and Medicine*, 90.2 (1991): 36–39; Colleen D. Clements and Roger C. Sider, "Medical Ethics' Assault on Medical Values," *Journal of the American Medical Association* 250 (1983): 2011.

[10]This is one of the themes Cheryl N. Noble raises in her influential paper, "Ethics and Experts," *Hastings Center Report* 12.3 (1982): 7–9.

[11]Françoise Baylis, A Profile of the Health Care Ethics Consultant, in this volume.

A Profile of the Health Care Ethics Consultant

Françoise E. Baylis

Preamble

This chapter provides a functional description of the health care ethics consultant who is active in the clinical setting. It outlines the requisite knowledge, abilities, and traits of character for case consultation on ethical issues in clinical care or in clinical research, and for ethics consultation to clinical ethics committees, research ethics boards (institutional review boards), and policy formulation committees in health care institutions. This chapter is the core of the book, the content of other chapters having been informed by the evolution of this chapter. This being said, a number of brief comments are necessary to place the chapter within its proper context.

First, it is important to note that individuals trained in applied ethics engage in a wide variety of activities, both within and outside of the clinical setting. The Profile outlined in this chapter is not for those involved in ethics education in an academic setting or policy making at a supra-institutional level. Although elements of this Profile may be relevant to the work of these applied ethicists, the Profile narrowly focuses on the activities of the health care ethics consultant in the clinical setting. Moreover, it assumes that the ethics consultant is actively engaged in all aspects of ethics consultation as described above. It follows that for those who are involved in a subset of these activities, certain elements of the Profile may not be relevant.[1]

Second, the knowledge and abilities listed in the Profile are minimum standards of practice, although clearly individual health care ethics consultants will have mastered the requisite knowledge and abilities to a greater or lesser extent, depending on upbringing, training, personal style, clinical experience, and institutional constraints. For this reason, a continuum from novice to expert is allowed for. More generally, the requirements of the Profile are not static; ongoing self-education and self-development are expected to ensure a current knowledge base and well-honed abilities.

Third, the virtues and traits of character listed in the Profile represent the ideal toward which the ethics consultant must strive. In addition to certain dispositions of the mind, effective ethics consultation requires certain dispositions of the heart, hence, the importance of virtue and character. Acknowledging human frailties, however, the virtues and traits of character are not proposed as minimum standards, but rather as standards to be pursued/practiced over a lifetime.

Fourth, the knowledge, abilities, virtues, and traits of character listed are neither exhaustive nor immutable. On reflection, there may be requirements listed that should be deleted, and others that should be added. Proposed changes to the Profile, however, should be the result of a process similar to the one that generated the Profile—from within and by consensus.

Fifth, the Profile of the health care ethics consultant is not a "stand-alone" document, and so to focus narrowly on the lists provided is to ignore the context within which the Profile was developed. This risks a serious misunderstanding of the initiative the Profile represents—a genuine effort to understand precisely what qualifies someone to be a health care ethics consultant.

The Profile was developed with a narrow focus on the role and responsibilities of the health care ethics consultant in the clinical setting. Of particular interest were the following questions: What should the ethics consultant know? What skills should s/he have? Who should s/he be? As far as possible, the Profile was developed without any starting assumptions about professional background or practice. In the spirit of a true quest for insight into an important question, the focus was on the end to be achieved (i.e., a functional description of the health care ethics consultant), and there were no assumptions made about the most appropriate means to the end. The discussion of "means to the end," found in the chapter on feeder disciplines, followed discussions of the Profile.

Finally, two brief comments are offered to those who will argue that the Profile sets the standard of practice far above that which can be achieved by the average person currently working in the clinical setting as a health care ethics consultant. First, we do not dispute the accuracy of this claim. Second, we do dispute the assumption that "what is, should be," particularly since in our view, there is a world of difference between a consultant and a dilettante, whether s/he works in ethics, medicine, engineering, architecture, and so forth. In our view, the Profile is an appropriate professional inventory of the knowledge, abilities, and traits of character required for effective health care ethics consultation.

Historical Context

A health care ethics consultant, at a minimum, is a person who helps to identify shared values by engaging others in ethical discourse as facilitator (e.g., mediator, negotiator) or other (e.g., confidant), and assists those who are confronted with ethically complex health care decisions in making choices that are morally acceptable to themselves and "within the bounds of communal and institutional acceptability."[2] It follows that an *effective* health care ethics consultant is someone who has the knowledge, abilities, and attributes of character to facilitate this type of ethical discourse in case consultation on ethical issues in clinical care or clinical research, and in ethics consultation to ethics committees, to research ethics boards (institutional review boards), and to policy formulation committees.

For Peter Singer, expertise in ethics requires a familiarity with moral concepts, an understanding of the logic of moral argumentation, and ample time to gather relevant information.[3] Essential for an ethical expert is:

> the ability to reason well and logically, to avoid errors in one's own arguments, and to detect fallacies when they occur in the arguments of others . . . an understanding of the nature of ethics and the meaning of moral concepts . . . reasonable knowledge of the major ethical theories . . . [and finally, knowledge of] the facts of the matter under discussion.[4]

For Arthur Caplan, this conception of moral expertise is too narrow. He is critical of the engineering model of applied ethics—a model that focuses exclusively on conceptual clarification, mastery of ethical theory, and impartiality—and he underlines the importance of moral diagnosis and moral judgment. In his view, to be effective in the health care setting, one must have the ability to identify and classify moral problems not previously discerned, and the ability to use moral knowledge to view moral problems from different perspectives. Caplan concedes, however, that in appropriate circumstances "engineering is a valuable and helpful activity even in a field such as ethics."[5] That is, at times the engineering model of applied ethics is a useful art that "requires practical knowledge, theoretical understanding, and experience."[6]

Terrence Ackerman accepts Caplan's general critique of the engineering model of applied ethics, and explicitly endorses the claim that the ethicist must be skilled in the classification and diagnosis of moral problems. Ackerman then adds that the primary role of the ethicist in the health care setting is to *facilitate* the reflective process

> by clarifying relevant moral values, conveying significant factual information, identifying alternative solutions, comparing the moral consequences of adopting these alternatives, and making recommendations for resolving the moral problem.[7]

These abilities, according to Ackerman, require:

> Knowledge of the purpose and process of moral reflection, familiarity with major moral principles and the historical source of their development, and skill in logical analysis of moral problems . . . current knowledge of the literature of bioethics . . . basic knowledge of medicine and medical terminology . . . [and] basic knowledge of the psychosocial literature relevant to moral issues in clinical care.[8]

The debate continues with Jonathan Moreno. In his article, "Ethics Consultation as Moral Engagement," Moreno provides what is perhaps the most recent comprehensive account of the effective health care ethics consultant. In addition to ethical expertise, which "involves at least (1) the knowledge of general principles and theories of morality, (2) analytic skills such as discernment and insight, and (3) the strength of will not to take the easy way out,"[9] the health care ethics consultant must have the following "other sorts of skill":

> First, the ethicist should be a skilled participant-observer, able to identify informal social structures and arrangements and to assess his or her developing role in them. Second, the ethicist should understand the dynamics of small group behavior, with an ability to recognize the interplay between sociometric structures and decisional outcomes. Third, the ethicist should be a competent mediator, familiar with negotiating strategies and having sound interpersonal skills.[10]

Moreover, "the non-physician ethicist must be familiar with the language of health care in order to be effective."[11]

In a subsequent article, published in the same year, "Call Me Doctor? Confessions of a Hospital Philosopher," Moreno adds:

> One must be acquainted with relevant statutory and case law, the institutional structure of the health care system, the financing of health care, and the prevailing consensus and current issues in health policy. Some understanding of the economies of health care is very useful and an appreciation for the sociological and political processes of the clinical setting is essential. Finally, sound interpersonal skills, particularly tactfulness and the ability to mediate among deeply felt differences while honoring them, can vastly enhance the value of the ethics consultant.[12]

Clearly, Moreno's description of the ethics consultant improves significantly on earlier descriptions provided by Singer,[13] Caplan,[14] Ackerman,[15] and others. Further refinement is nonetheless possible, and the following functional description of the health care ethics consultant is offered in that spirit. This Profile lists knowledge, abilities, and traits of character for case consultation (clinical or research), committee consultation (clinical or research), and policy formulation.

Knowledge Requirements

1. Extensive knowledge not only of the current health care ethics literature,[16] but also of "classic" articles and influential cases in health care ethics.
2. Extensive knowledge and critical understanding of at least one ethical theory/tradition/cultural belief system that is rich enough to allow one to develop a style of thought, habits of rigor, and judgment.[17]
3. Extensive knowledge and critical understanding of the concepts of "health," "illness," "clinical practice," and "medical research" in the health care system in which one is working (e.g., Western medicine, a comprehensive government-funded health care system).
4. Extensive knowledge of one's own biases/partiality.
5. Extensive knowledge of facilitation (e.g., mediation, negotiation, and arbitration) techniques, and knowledge of underlying theory.
6. Knowledge of the following:
   a. Medical terminology.[18]
   b. Common health care problems.
   c. Emerging health care problems.
   d. The range of health care settings.
   e. The strengths and limitations of the scientific method and the medical model of health care.
   f. The health care system's structures and decision making methods.
   g. Relevant institutional ethos and policies.
   h. Relevant professional guidelines[19] and codes of ethics.
7. Knowledge of various ethical theories/traditions/cultural belief systems that are most commonly held by health practitioners, patients, families, administrators, and social agencies in the health care system in which one is working.
8. Knowledge of the human dimension(s) of ethical problem-solving which includes an understanding of the social and cultural circumstances that affect the patients' and caregivers' emotional responses to a health problem.

9. Knowledge of health law, including knowledge of relevant government regulations, policy statements, legislation, and legal cases.[20]
10. Knowledge of cultural differences relevant to beliefs about health care.
11. Knowledge of available resources (e.g., community support systems, contact persons, and national organizations).

To elaborate briefly on the knowledge requirements listed above, it is important to note that these requirements are essentially summary statements. Consider, for example, knowledge requirement number 6 (K6). It is a succinct reference to the subject areas in medicine that a health care ethics consultant should be familiar with. A somewhat expanded version of this requirement would include the following:

1. Knowledge of medical terminology: An effective health care ethics consultant should understand a wide range of medical terms from such basic terms as diagnosis, prognosis, benign, malignant, fetus or embryo, to more complex terms, such as multiorgan or multisystem failure, coma, brain-stem function, persistent vegetative state, and brain death. However, the health care ethics consultant would not be expected to have knowledge of less common, more esoteric terms, such as panhypopituitarism or holoprosencephaly, if s/he had never been involved in a case where these diagnoses were relevant.
2. Common health care problems: An effective health care ethics consultant should appreciate the clinical dimensions of common diseases, for example, the etiology, medical work-up, diagnosis, course of illness, prognosis, and available interventions for lung cancer. By contrast, the ethics consultant need not be as familiar with Jakob-Creutzfeldt disease; s/he need only know how to find and select relevant information when this is missing.

3. Emerging health care problems: Although the health care ethics consultant could not be expected to be aware of recent developments in every branch of medicine, s/he should become familiar with health care issues as soon as they are recognized to have potentially significant implications—either because of the extent of their impact, or because of potential social and political ramifications (e.g., Resource Allocation, Acquired Immunodeficiency Syndrome [AIDS], and Assisted Reproductive Technologies).

Second, although not listed strictly in order of importance, the knowledge requirements fall into two distinct broad categories. First, "extensive knowledge" of facilitation techniques, self, fundamental concepts in health care, ethics, and health care ethics is required—these areas of knowledge being at the core of health care ethics consultation. For the rest, only practical "working" knowledge—actual or potential—is needed. Potential working knowledge presumes the ability to become knowledgeable "enough" for the task at hand. In any case, the value of the knowledge (whether extensive knowledge or working knowledge) lies in the ethics consultant's ability to use the knowledge.

Requisite Abilities

1. The ability to identify ethical issues coupled with the ability to sensitize others to ethical issues.
2. The ability to acquire relevant information (e.g., clinical information, psychosocial information) when gaps are revealed, coupled with the ability to know which questions to ask when attempting to fill in the gaps.
3. The ability to bring systematic thinking to ethical problem-solving.[21]
   a. The ability to analyze the meanings of concepts and principles.[22] Such analysis should include a

consideration of the meaning of words in casual everyday contexts, their technical meaning, and their idiosyncratic use.

b. The ability to identify underlying assumptions[23] and to question those that are believed to be flawed or misleading.[24]
c. The ability to reframe a presenting ethical problem when appropriate (i.e., not in an attempt to evade or to distract), and thereby to shift the ethical discussion so as to allow for a consideration of alternative options that might otherwise not be considered.[25]

4. The ability to make and defend sound ethical judgments that reflect an understanding of the values of others, including:
   a. The ability to identify possible alternative courses of action,[26] to outline the associated values and possible consequences,[27] and to provide the best arguments for and against the various options; and
   b. The ability to provide a recommendation[28] "for consideration,"[29] without attempting to manipulate the decision making process.[30]
5. The ability to communicate effectively with (which includes actively listen to) health care workers, patients, patients' families, administrators, and social agencies.[31]
   a. The ability to elicit, appreciate, explore, and (when necessary) help clarify[32] the evolving/changing viewpoints, beliefs, and values of others.
   b. The ability to represent the evolving/changing viewpoints, beliefs, and values of one party to another.
   c. The ability to observe and perceive interests and relationships that influence discussion and behavior.[33]
6. The ability to facilitate (e.g., mediate, negotiate, and arbitrate), coupled with an ability to ascertain when one or more of these activities is appropriate.[34]

7. The ability to recognize and work within the limits of one's knowledge balanced with the ability to accept challenges.
8. The ability to recognize one's own partiality and not to introduce personal beliefs (i.e., values or traditions one finds persuasive) in an inappropriate manner. Specifically, personal beliefs and value commitments should be identified in a timely fashion when there is an overriding commitment to a specific value system that is informing the discussion (e.g., religious or political beliefs). Personal beliefs and value commitments should not be introduced: (a) in a clandestine or subliminal way, or (b) under false colors—as facts when they are not, as consensually agreed on when they are not, or as legal requirements when they are not.
9. The ability to participate in group decision making, even when this may generate a conclusion with which one disagrees, and a willingness to tolerate such group decisions. This ability must be coupled with a strong sense of personal and professional integrity so that one may distinguish outcomes that one will not, on moral grounds, endorse/sanction (e.g., an outcome that violates an important moral principle, but is chosen to avoid confrontation) from outcomes that one disagrees with, but will endorse/sanction (e.g., an outcome within communal and institutional norms).
10. The ability to withstand the influence of public opinion and to question existing traditions, customs, and laws.

To be sure, different clinical settings and different clinical situations necessarily require the exercise of different abilities. The Profile attempts to provide a reasonably complete listing of the minimum skills one should be able to draw on, as necessary. It is not expected, nor for that matter would it be productive, for an ethics consultant to exercise all of these abilities at any one time. Consider, for example, the ability to facilitate (e.g., mediate, negotiate, and arbitrate) (A6). Clearly, different facilitation roles are appropriate for different activities in different circumstances.

Traits of Character

Having outlined the requisite knowledge and abilities of the health care ethics consultant, we now turn our attention to a most critical question: What kind of person should the health care ethics consultant be? That is, what traits of character, what virtues, and what attitudes should s/he have to be effective in the clinical setting?

A survey of the literature on health care ethics consultation reveals an interesting and recent shift in the description of the ethics consultant. The early writings focus exclusively on knowledge and abilities. Occasional mention is made of the need for "detachment,"[35] but this character trait is generally described as an ability (e.g., the ability to remain detached, disinterested, neutral, and objective). Similarly, although there are references to "trust," these are not references to trustworthiness, but to the ability to inspire trust and confidence.[36] Also, instead of extolling the virtue of empathy, the claim is that the ethics consultant should "have reasonable skills in expressing empathy."[37]

James Drane offers a plausible explanation of this tendency to reduce character and virtue to abilities. He posits that:

> The very words, virtue and character, have a religious ring to the secular thinker, which is reason enough to consider them out of place. To be seriously considered, these words would have to be "laundered" and "operationalized."[38]

Hence, writes Howard Brody,

> the tendency among analytic philosophers of medical ethics to reduce virtue and character to tendencies to behave or decide in certain ways, and to miss

> the point that these concepts presuppose standards of excellence, rather than rules for minimally acceptable behavior, and the coherence of one's actions over a lifetime, rather than discrete decisions.[39]

Ironically, particularly with the earlier writings, instead of references to the character of the health care ethics consultant, there are references to the character of prospective employers of ethics consultants. Caplan, for one, writes that independence and tolerance are required of those who would engage the ethics consultant.[40]

More recently, however, the relevance of virtue and character has been emphasized. Arthur Schafer, for example, writes in the lay press of the ethicist's need for a sense of humility.[41] John Robertson writes of the duty to be "reasonably humble."[41] Françoise Baylis writes on the importance of patience, tolerance, empathy, and intellectual honesty.[43] George Agich speaks of the need for gentleness.[44] David Thomasma emphasizes the importance of prudence and fortitude, and decries pusillanimity as a vice.[45] Moreno lauds the importance of "strength of will not to take the easy way out."[46]

These limited references to virtue and character aside, the critical question remains: What sort of person should the ethics consultant be? First and foremost, the health care ethics consultant needs wisdom to reason through the stages of deliberation and judgment toward ethically defensible options, recommendations, and actions. "Good intentions" and "meaning well" are insufficient; nothing less than open-minded, clearsightedness, and foresight will suffice. The health care ethics consultant needs wisdom to reach and to evaluate the goals of ethics consultation.

Second, justice is important in order to secure the cooperation of, and mutual trust among, health care professionals, patients, patients' families, and others. Justice is the virtue that directs the ethics consultant in her/his dealings with others and ensures that s/he gives to the other that which is her/his due. It presumes impartiality, fairness, and honesty.

In addition to wisdom and justice, the health care ethics consultant needs courage. In the clinical setting, when faced with a serious wrong, the ethics consultant may "lack the strength of will not to take the easy way out." The personal and professional consequences of calling into question the moral acceptability of certain practices, decisions, institutional structures, or policies could be significant. Furthermore, there will be tremendous pressure to conform so as to gain the approval of one's colleagues in the health care setting. An effective health care ethics consultant, however, is not a compliant colleague. S/he needs courage to take a stand in the face of serious wrong, and courage to persevere in the face of seemingly constant "setbacks, weariness, difficulties, and dangers."[47] Courage inclines the ethics consultant away from a self-centered concern for security.

In addition, the health care ethics consultant needs temperance to avoid being distracted from worthy long-term goals by short-term pleasures. Now typically, temperance refers to the need for moderation in eating, drinking, and the pursuit of sensual pleasures. In this instance, however, the reference is more broadly to the need for self-control. As Laura Purdy argues:

> some measure of self-control is an essential accompaniment of virtually everybody's life . . . With self-control, we can lay down the basis for the enabling virtues necessary if people are both to pursue their own self-interest and to further the welfare of others.[48]

With ethics consultation, for example, there is a need for moderation in the pursuit of fortune and fame. A risk for the few who practice as health care ethics consultants is that they will be tempted to pursue personal interests over and above other interests they ought to promote in their capacity as health care ethics consultant. There is, for example, the lure of television, the court room, the podium, the international lecture circuit, and so on. To be sure, the pursuit of money and recognition (importance and status) is not inherently wrong, and in some cases, the associated prestige may well enhance one's effectiveness in the clinical setting. However, the health care ethics consultant must be wary of sacrificing personal and professional integrity to her/his desire for fortune and fame.

To the virtues briefly discussed above,[49] compassion, humility, and integrity can be added. The health care ethics consultant should be compassionate and caring—able to appreciate the needs of, and show genuine concern for, those who are suffering. S/he should not distance her/himself from the emotional aspects of the presenting ethical problem, but should strive to understand the anguish of those who are suffering, and their need for comfort and help.

Also, there is the virtue of humility, which inclines the ethics consultant away from the temptation to think too well of her/himself. The health care ethics consultant should be humble in acknowledging her/his limits. Ideally, genuine self-knowledge yields self-critical humility, i.e., an "awareness of one's own limitations and need for others . . . that ultimately results in a sense of equality with all others."[50] The health care ethics consultant should also have integrity—an uncompromising commitment to be honest and trustworthy in the exercise of one's functions as an ethics consultant.

From the virtues described above, a number of other enabling virtues might be derived. For example, one might argue that justice requires veracity and fidelity (telling the truth and keeping promises); courage requires endurance and patience; and compassion requires kindness and charity. However, the purpose here is not to provide an exhaustive list of the virtues that a good ethics consultant must possess, but rather to emphasize the importance of virtue and character for effective health care ethics consultation.

The point is that knowledge and abilities are not sufficient, particularly since the abilities listed in the Profile presume certain virtues. For example, the ability to make and defend sound ethical judgments requires wisdom, as does the ability to distinguish outcomes that, for moral reasons, one will not endorse from outcomes that one disagrees with, but will endorse. Likewise, the ability to participate in group decision making even when this may generate a conclusion with which one disagrees, and the willingness to tolerate such a decision, require justice. Furthermore, the ability to bring systematic thinking to ethical problem-solving requires courage given that "critical thinking is almost pointless unless one is prepared to act in accordance with it."[51] More generally, it takes a courageous person to make a difference. The ability to recognize and work within the limits of one's knowledge requires humility. The ability to provide a recommendation without attempting to manipulate the decision making process requires integrity, and so on.

Conclusion

In closing, no doubt many will object to the Profile of the health care ethics consultant and, in particular, we (the Network members) anticipate that most objections will suggest that the elements of the Profile are too onerous. Such objections were certainly raised by audiences to whom drafts of the Profile were presented. In our view, however, no elements of the Profile are peripheral or superfluous to effective ethics consultation. To return to our opening remarks, however, there is no claim made here to the effect that the knowledge, abilities, virtues, and traits of character listed in the Profile are either exhaustive or immutable. The Profile is our best approximation of what is required for effective health care ethics consultation, and we invite those who share our view about effective ethics consultation to improve on the Profile. The challenge we put to the doubting reader is to develop and defend a cogent argument for removing any one of the elements of the Profile.

Notes and References

[1]Note, in determining which elements of the Profile are relevant, the issue is not the amount of time spent in health care ethics consultation (e.g., full-time vs part-time), but rather the breadth of activities one is engaged in.

[2]Erich H. Loewy, "Ethics Consultation and Ethics Committees," *HEC Forum* 2 (1990): 357.

[3]Peter A. Singer, "Moral Experts," *Analysis* 32.4 (1972): 116,117.

[4]Peter A. Singer and Deane Wells, *Reproduction Revolution: New Ways of Making Babies* (Oxford: Oxford University Press, 1984) 200.

[5]Arthur L. Caplan, "Can Applied Ethics Be Effective in Health Care and Should It Strive to Be?" *Ethics* 93 (1983): 319.

[6]Caplan, "Applied Ethics" 319.

[7]Terrence F. Ackerman, "The Role of an Ethicist in Health Care," *Health Care Ethics: A Guide for Decision Makers*, eds. Gary R. Anderson and Valerie A. Glesnes-Anderson (Rockville, MD: Aspen, 1987) 315.

[8]Ackerman, "Role" 317.

[9]Jonathan D. Moreno, "Ethics Consultation as Moral Engagement," *Bioethics* 5.1 (1991): 47.

[10]Moreno, "Ethics Consultation" 55.

[11]Moreno, "Ethics Consultation" 55.

[12]Jonathan D. Moreno, "Call Me Doctor? Confessions of a Hospital Philosopher," *Journal of Medical Humanities* 12.4 (1991): 194,195.

[13]Singer 116–7; Singer and Wells 200.

[14]Arthur L. Caplan, "Ethical Engineers Need Not Apply: The State of Applied Ethics Today," *Science, Technology and Human Values* 6.33 (1980): 24–32; Arthur L. Caplan, "Mechanics on Duty: The Limitations of a Technical Definition of Moral Expertise for Work in Applied Ethics," *New Essays in Ethics and Public Policy*, eds. Kai Nielsen and Steven C. Patten, supplement VIII

to *Canadian Journal of Philosophy* (1982): 1–18; Caplan, "Applied Ethics" 311–319; and Arthur L. Caplan, "Moral Experts and Moral Expertise: Do Either Exist?" *Clinical Ethics: Theory and Practice*, eds. Barry Hoffmaster, Benjamin Freedman, and Gwen Fraser (Clifton, NJ: Humana, 1989) 59–87.

[15]Ackerman, "Role" 308–320; Terrence F. Ackerman, "Moral Problems, Moral Inquiry, and Consultation in Clinical Ethics," *Clinical Ethics: Theory and Practice*, eds. Barry Hoffmaster, Benjamin Freedman, and Gwen Fraser (Clifton, NJ: Humana, 1989) 144–160.

[16]Cf Ackerman, "Role" 317; Albert R. Jonsen, "Can an Ethicist Be a Consultant?" *Frontiers in Medical Ethics: Applications in a Medical Setting*, ed. Virginia Abernathy (Cambridge: Balinger, 1980) 157; Jacqueline J. Glover, David T. Ozar, and David C. Thomasma, "Teaching Ethics on Rounds: The Ethicist as Teacher, Consultant, and Decision Maker," *Theoretical Medicine* 7 (1986): 15,23.

[17]E.g., "knowledge of a process of moral reasoning." Glover 15.

[18]E.g., "the non-physician ethicist must be familiar with the language of health care in order to be effective." Moreno, "Ethics Consultation" 55; "basic knowledge of medicine and medical terminology is an important requirement." Ackerman, "Role" 317.

[19]Cf Ackerman, "Role" 314.

[20]Cf Glover 19, 23; Ackerman, "Role" 314.

[21]E.g., "good quality reasoning is of crucial importance." E. Haavi Morreim, "Philosophy Lessons from the Clinical Setting: Seven Sayings That Used to Annoy Me," *Theoretical Medicine* 7 (1986): 55.

[22]E.g., "In its narrow sense, to do ethics is to be good at doing what well-trained philosophers and theologians do: analyze concepts, clarify principles, see logical entailments, spot underlying assumptions, and build theoretical systems." Daniel Callahan, "Bioethics as a Discipline," *Hastings Center Studies* 1.1 (1973): 73.

[23]E.g., "to quickly recognize the presuppositions, flawed or valid, hiding behind first order judgments." Benjamin Freedman, "One Philosopher's Experience on an Ethics Committee," *Hastings Center Report* 11.2 (1981): 22; "he can expose the implicit assumptions which are shaping the physician's description of the situation." Morreim 55.

[24]E.g., "a philosopher can at least help to call false assumptions to attention." Morreim 59; an ability "to clarify options by noting questionable assumptions or neglected moral considerations

that tip the scales." William Ruddick, "Can Doctors and Philosophers Work Together?" *Hastings Center Report* 11.2 (1981): 16.

[25]E.g., "The applied ethicist must feel free to reinterpret complaints, disregard some issues, and occasionally, move beyond the issues as initially framed." Caplan, "Applied Ethics" 317.

[26]E.g., "the consulting ethicist can provide assistance in identifying alternative plans of action." Ackerman, "Moral Problems" 154.

[27]E.g., "to outline advantages and disadvantages of various options." Loewy 353.

[28]E.g., "Ethicists are expected to take a position, give advice, express an educated opinion or at the very least offer constructive options." Moreno, "Ethics Consultation" 52.

[29]E.g., "the consulting ethicist is not able to deliver "right answers" to moral problems, but may provide informed recommendations for consideration." Ackerman, "Moral Problems" 157.

[30]E.g., "The most important point about making recommendations in clinical cases is that the recommendation does not represent an imposition of values on others; rather it flows from the values of the patients and physicians, and other caregivers, examined in the case." David C. Thomasma, "Why Philosophers Should Offer Ethics Consultations," *Theoretical Medicine* 12 (1991): 137.

[31]Freedman 22.

[32]E.g., "value clarification." Gregory T. Lyon-Loftus, "What is a Clinical Ethicist?" *Theoretical Medicine* 7 (1986): 42.

[33]E.g., "The ethicist should understand the dynamics of small group behavior, with an ability to recognize the interplay between sociometric structures and decisional outcomes." Moreno, "Ethics Consultation" 55.

[34]E.g., "a facilitator of the meeting, guiding the discussion, helping to gather important information, and helping the group reach a consensus among individuals representing varied interests." Glover 19; "the basic function of the ethics consultant is to *facilitate* the process by which reflective resolution of a moral problem may be achieved." Ackerman, "Moral Problems" 153.

[35]E.g., Ruddick 16,17.

[36]E.g., Ackerman, "Role" 318.

[37]Joel Frader, "Political and Interpersonal Aspects of Ethics Consultation," *Theoretical Medicine* 13 (1992): 31–44.

[38]James F. Drane, *Becoming a Good Doctor: The Place of Virtue and Character in Medical Ethics* (Kansas City: Sheed, 1988) 142.

[39]Howard Brody, *The Healer's Power* (New Haven: Yale University Press, 1992) 253.

[40]Caplan, "Applied Ethics" 319.

[41]Arthur Schafer, "Are Hospital Ethicists Doing a Worthwhile Job?" *Globe and Mail* Dec. 9, 1986: A7.

[42]John A. Robertson, "Clinical Medical Ethics and the Law: The Rights and Duties of Ethics Consultants," *Ethics Consultation in Health Care*, eds. John C. Fletcher, Norman Quist, and Albert R. Jonsen (Ann Arbor: Health Administration, 1989) 166.

[43]Françoise Baylis, "Moral Experts and Moral Expertise: Wherein Lies the Difference?" *Clinical Ethics: Theory and Practice*, eds. Barry Hoffmaster, Benjamin Freedman, and Gwen Fraser (Clifton, NJ: Humana, 1989) 96–97.

[44]George Agich, "Clinical Ethics: A Role Theoretic Look," *Social Science and Medicine* 30.4 (1990): 396.

[45]Thomasma 129–40.

[46]Moreno, "Ethics Consultation" 47.

[47]Peter Geach, *The Virtues* (London: Cambridge University Press, 1977) 16.

[48]Laura Purdy, *In Their Best Interest? The Case Against Equal Rights for Children* (Ithaca, NY: Cornell University Press, 1992) 47,48.

[49] The initial list of virtues (wisdom, justice, courage, and temperance), but not their application, is borrowed from Aristotle, *Nichomachean Ethics*.

[50]Karen Lebacqz, "Humility in Health Care," *Journal of Medicine and Philosophy* 17 (1992): 299,300.

[51]Purdy 62.

" . . . Has Knowledge of [Interpersonal] Facilitation Techniques and Theory; Has the Ability to Facilitate [Interpersonally] . . . "*

Fact or Fiction?

Abbyann Lynch

Various authors, including members of the Network, have identified facilitation as one of several forms of knowledge (K5) and abilities (A6) essential to the sound pursuit of the health care ethics consultation process.[1] In this context, facilitation (an individual's ability to make easy, to promote, to help forward an action or result) can be understood to comprise at least two quite different, but complementary activities.

In the first instance, the health care ethics consultant is said to act as facilitator when contributing to

> the classification and diagnosis of the moral problem . . . [when providing] assistance in identifying alternative plans of action . . . [when promoting] a

*From pp. 31 and 34.

> clearer understanding of the factual components of a morally problematic situation . . . [when] assessing how alternative plans of action will promote or impede the realization of various valued states of affairs . . . [when making] recommendations regarding the proper resolution of moral problems. This role . . . as a facilitator of moral reflection can be implemented in various activities . . . staff education, policy formation, and clinical case consultation.[2]

Such facilitation is identified here as "intellectual facilitation."

Facilitation can also be understood as that knowledge and ability demonstrated by the health care ethics consultant when the theory and skills relevant to the use of interpersonal communication and group dynamics are employed in the service of a larger or smaller group's activities of moral reflection and dilemma resolution. Thus, for example, the health care ethics consultant acts as "a facilitator of the meeting, guiding the discussion . . . helping the group reach a consensus among individuals representing varied interests";[3] the health care ethics consultant:

> understand[s] the dynamics of small group behaviour . . . [has] an ability to recognize the interplay between sociometric structures and decisional outcomes . . . [is] a competent mediator, familiar with negotiating strategies and [has] sound interpersonal skills;[4]

the ethics consultant's special skills include the ability to "negotiate and facilitate negotiations."[5] Such facilitation is identified here as "interpersonal facilitation"; it allows "intellectual facilitation" to make its appropriate contribution in the immediate situation.

Interpersonal facilitation in health care ethics consultation will be analogous to its use in various other human endeavors, including business (labor negotiation, arbitration), law (adjudication), marriage counseling (conflict resolution), and diplomacy (negotiation). In each of these activities, the desired goal can be achieved only by the use of some form of interpersonal relationship; such a relationship allows communication of certain knowledge (concerning business, law, and so on) relevant to the discussion at hand. Such interpersonal facilitation can be understood as the more general rubric under which such activities as problem-solving, refereeing, bargaining, and compromising can be grouped. The core knowledge requisite to the development of skill in the pursuit of such activities is drawn from the discipline of psychology, particularly social psychology. Such knowledge is applied using specialized techniques, such as clarification, listening, open communication, role-playing, persuasion, confrontation, harmonizing, repetition, suggestion, and creation of options.

In the context of health care ethics consultation, facilitation, as a type of communication-of-knowledge methodology, comprises some form of dialogue with others ("interpersonal facilitation") for the purpose of applying various theoretical knowledges ("intellectual facilitation") toward the resolution of a presenting ethical challenge. As a subset of such facilitation, interpersonal facilitation may be seen as the means employed to bring theoretical concepts to the attention of the clinical ethics committee that is considering a particular case; it may enable the research ethics committee to continue the discussion needed to reach ethical consensus regarding a particular protocol; it might encourage members of a policy formulation committee to give more careful attention to the ethical dimension of their work. Such interpersonal facilitation skill, when exercised in a smaller group of individuals who "own" a case,[6] but who differ about its resolution, can help keep those individuals communicating with each other. This process may also assist them in some joint ethics reflection, perhaps leading toward appropriate compromise regarding their difference(s).

Interpersonal Facilitation Modalities Used in Health Care Ethics Consultation

In the course of health care ethics consultation, interpersonal facilitation is employed in at least three ways—negotiation, adjudication, and mediation.[7] The use of any one of these facilitation modalities in place of another will depend on a number of factors, including the type of problem presented and the individuals involved in it; the institution in which the ethics consultation service is offered and its ethos; and the ethics consultant's assigned role and personal style.

When acting as a negotiator, the health care ethics consultant may serve as an enabler—aiding the individuals who "own" the dilemma, but who differ about its resolution, to work toward a compromise position that is agreeable to themselves. In this variety of negotiation, the ethics consultant may clear away obstacles to communication, clarify terms, or frame options. Actual reconciliation of ethical differences is left to those concerned and desirous of achieving reconciliation.[8] Such a negotiating posture is evidently quite different from that adopted by the health care ethics consultant who negotiates on behalf of others. In this type of negotiation, the ethics consultant will be acting more directly, whether at the request of an individual or a group or on personal initiative, and whether individually or as one of a group.[9] In such activity, the ethics consultant will be immediately involved in a type of bargaining discussion, serving as a direct spokesperson for others or for oneself; the goal may be the initiation of a change in policy or the reconsideration of a treatment decision. In either type of negotiating, the opportunity that is created, recognized, or seized on to bring appropriate ethical considerations to the attention of those involved will depend greatly on the ethics consultant's interpersonal facilitation skills.

The health care ethics consultant may also serve as adjudicator or arbitrator. When asked for a determining opinion with reference to the ethical appropriateness of an institutional policy, for example, the consultant may be acting more as an umpire or referee. In this situation, presuming the opinion is final and not subject to petition for reconsideration, the ethics consultant stands apart from others with an interest in the matter, providing the "ultimate word," as in the court of last resort. Such skilled actiivity is different in kind from that used by the ethics consultant acting as equal to peers in a group that is discussing, for example, the modification of a research consent form. Such adjudication skill differs as well from that of the negotiator, who is expected to act as advocate, first bargaining, and then, perhaps, compromising in the pursuit of a particular value goal regarding a particular case. When serving as an adjudicator, the ethics consultant must gain the trust and respect of everyone involved, first by listening to them and then by impartially discerning what is the best action to take, all things considered.

In addition to the role of negotiator and adjudicator is the role of mediator, and experience suggests that mediation is the facilitation modality most frequently employed by the health care ethics consultant. In this activity, assistance is sought in the resolution of an identified dilemma by the persons who "own" the dilemma, and the consultant is asked to act as a link between/among them in sorting out differences in viewpoint.[10] In this mediating capacity, the ethics consultant may intervene directly in any discussions among those concerned by using the various techniques mentioned, and by "forcing" or guiding those responsible for decision making in the particular instance to achieve some kind of (desired) unity or "settlement" between/among the opposing points of view.

As employed in the context of health care ethics consultation, resort to any of these activities (negotiation, adjudication, mediation) is a voluntary activity; such conclusion as may be reached is not necessarily legally enforceable. In these and other ways, facilitation in ethics consultation may differ significantly from facilitation as practiced in business, public service, or the reconciliation of domestic disputes.

Process and Structure of Interpersonal Facilitation in Health Care Ethics Consultation

Literature regarding the use of negotiation, adjudication, and mediation skills in a variety of settings (business, labor, marriage counseling, and so forth) abounds. There is also an abundance of literature regarding the ethics consultant's use of "intellectual facilitation" in the resolution of an ethical dilemma, whether in reference to business, environmental, or journalistic pursuits. Surprisingly, however, there is little comment in the literature regarding the ethics consultant's knowledge and use of "interpersonal facilitation" in these various contexts, much less in the context of health care.

A notable exception is an article by Mary Beth West and Joan McIver Gibson,[11] in which they attempt to adapt a legal understanding of the processes of mediation (assistance in settlement of defined disputes) and facilitation (a term they use when speaking of assistance where a non-defined dispute/difference appears to exist) to the activity of the ethics committee involved in case review. In so doing, the authors list activities they regard as common to these processes: intake; introduction and contracting; information gathering and issue identification; agenda setting and reaching resolution; and agreement writing and follow-up.[12] Supplementary to this listing, the authors offer further helpful comments regarding appropriate actions at each of these stages, again as applied in the context of ethics committee consultation. Speaking of intake, for example, they note the need for identification of interested parties, introduction, making each party comfortable, and establishing rapport; gathering information and maintaining an open, sensitive climate; problem clarification and summary; review of possible processes for addressing the issue; and selection of the option.[13] The various stages and ancillary activities noted by West and Gibson are evidently common to many other forms of conflict resolution. They are readily recognized as suitably complementary to the approaches used by the ethicist in "theoretical" resolution of an ethical dilemma (i.e., the process of intellectual facilitation regarding, for example, identification of the presenting ethical conflict, fact finding, consideration of relevant values, development of options for action, and consideration of consequences).

Ethical Interpersonal Facilitation in the Practice of Health Care Ethics Consultation

Able use of the processes involved in the health care ethics consultant's practice of negotiation, adjudication, or mediation as modes of interpersonal facilitation rests on personal skill in the use of certain psychological techniques already noted. Any more specific discussion regarding the identification and application of such techniques is beyond the scope of this chapter. However, some further comment is in order as to the "good" use of these relevant techniques.

Whatever form of interpersonal facilitation is used, such use may be identified as "ethical" or not. Thus, for example, the ethics consultant who is expert in eliciting discussion among peers or among those who differ about resolution of a particular case may behave in a scurrilous fashion, using such disclosure for personal advantage. In this respect, the more general facilitation literature provides preliminary guidance; its recommendations are consistent with many found in the literature on ethics consultation, and with the observation of Françoise Baylis: Knowledge and abilities are not sufficient . . . virtue and character are important for effective health care ethics consultation.[14]

As examples of "good" practice in the use of mediation, consider the following. Jay Folberg and Alison Taylor warn against "muscle mediators," i.e., those who "inform disputants of the best 'voluntary' resolution or who narrow the options to preclude effective choices."[15] Such a caveat is applicable as a reminder to the health care ethics consultant that ethical decisions regarding ethical dilemmas "owned" by others are usually to be made by the "owners"; the ethics consultant must thus be *respectful* in providing all necessary information to that end.

Closely related to this advice is that provided by Kent E. Menzel, who argues cogently regarding the need for *fairness* in such activity.[16] Adapted to the health care ethics consultation context, attention to fairness is a necessary consideration in attempting to balance power among the consultation's participants (for example, as in mediating the situation of ethical difference between the provider of care and its recipient, or between a child-patient and parents). West and Gibson reiterate this point, again adapting legal considerations to those of ethics case review, as they list various institutional and personal sources of power.[17] They provide an implicit warning as to the need for those engaged in ethics consultation to be aware of the asymmetry of power in mediation contexts; in so doing, they cite Bernard Mayer's work in mediation as a source for listing possible problem areas.[18] Thus, they mention formal authority (as in the case of a judge or Chief Executive Officer), expert power (derived from having certain expertise), associated power (derived from association with other people of power), resource power (derived from having control over valued resources, such as money or materials), procedural power (derived from having control over the procedures by which decisions are made), sanction power (derived from the ability, or the perceived ability, to inflict harm or to interfere with another's ability to realize his or her interests), nuisance power (derived from the ability to cause discomfort), habitual power (derived from the status quo), moral power (derived from an appeal to widely held values), and personal power (derived from a variety of personal attributes, such as self-assurance, determination, and endurance).[19]

The need for *neutrality* in the practice of legal mediation or facilitation (and thus of impartiality in their ethics consultation counterpart) is noted by West and Gibson.[20] A similar concern is expressed by Bethany June Spielman,[21] who addresses the use of mediation in ethics consultation, drawing attention in her analysis to the possible conflict of interest following from the often unrecognized normative bias of the ethicist engaged in problem-solving—a point frequently made in the ethics literature. Several authors identify the important role of the facilitator in establishing a level of *trust* in such consultation,[22] a point made by various authors in the ethics consultation literature, including members of the Network. Lacking trust in the consultant, those seeking assistance in the process of mediation, for example, could become frustrated and disillusioned. The ability of the consultant to maintain *confidentiality* is also highly valued;[23] indeed, quite beyond its value in the facilitation practiced by the health care ethics consultant, this virtue is identified as essential in all areas of health care.

Many other allusions to the "good" qualities expected in the facilitator (many of them equally necessary in the would-be ethics consultant) could be cited. A longer listing of desirable qualities of a mediator (more generally, a facilitator) is offered by Howard Raiffa, quoting William Simkin:

> In a jocular mood, [Simkin] wrote that a mediator should have:
>
> 1. The patience of Job;
> 2. The sincerity and bulldog characteristics of the English;
> 3. The wit of the Irish;
> 4. The physical endurance of the marathon runner;
> 5. The broken-field dodging ability of a halfback;
> 6. The guile of Machiavelli;
> 7. The personality-probing skills of a good psychiatrist;
> 8. The confidence-retaining characteristics of a mute;
> 9. The hide of a rhinoceros; and
> 10. The wisdom of Solomon.
>
> And, in a more serious vein, he added the following:
>
> 11. Demonstrated integrity and impartiality;
> 12. A basic knowledge of and belief in the collective bargaining process;
> 13. Firm faith in voluntarism, in contrast to dictation;
> 14. A fundamental belief in human values and potentials, tempered by the ability to assess personal weaknesses as well as strengths;
> 15. A hard-nosed ability to analyze what is available, in contrast to what may be desirable; and
> 16. Sufficient personal drive and ego, qualified by a willingness to be self-effacing.[24]

Education for the Practice of Interpersonal Facilitation in Health Care Ethics Consultation

It is generally recognized that appropriate application of the psychological techniques used in the practice of interpersonal facilitation can be learned, and should be perfected through supervised practice and repeated use. Folberg and Taylor, for example, explore several of the opportunities available in mediation education and training.[25] The spectrum mentioned ranges from training by way of one-day workshops, through one-week training programs, to degree or certificate curricula in mediation, or conflict resolution as offered at the University of Colorado, Catholic University of America, the University of Maryland, and the University of Illinois.

Folberg and Taylor also identify several specific content areas to be mastered by those who would consider themselves capable of serving as mediators. They must understand:

> the nature of conflict, how it comes into being, its dynamics, how it is manifested in particular disputes, and how it may be managed and resolved. . . [as well as mediation procedure and the importance of] disassociation of habits and assumptions fostered by the mediator's original profession.[26]

These individuals must have knowledge and ability regarding a variety of skills particular to mediation, as well as substantive knowledge regarding the discipline in question, for example, ethics, law, labor relations, and accounting.[27]

Worthy of particular comment here is Folberg and Taylor's warning that the would-be mediator must acquire a "change of internal map," i.e., effect some distance or disassociation from the mediator's original profession. For example, they argue that the lawyer engaged in mediation must set aside the assumption that the

> disputants are adversaries and that disputes are resolved by the proper application of legal rules. [Similarly, they note that the] . . . therapist's philosophical map, which emphasizes finding and treating the causes of conflict while allowing individual

decisions to be deferred, must be converted to the joint decisions and task orientation of mediation.[28]

What are the implications of Folberg and Taylor's warning for the practice of mediation by the health care ethics consultant? Clearly, the health care ethics consultant cannot function as philosopher only (or physician only, or lawyer only) when engaged in this activity. The philosopher (or physician or lawyer) who would act as mediator in the matter of health care ethics brings to this activity many kinds of knowledge and a variety of skills useful in the practice of mediation. Still, if the health care ethics consultant is to achieve the goal of ethics consultation using mediation, for example, there must be a conscious effort to merge his or her professional knowledge and skills with the knowledge and skills of both mediation and ethics (if ethics is not her first discipline).

To argue the need for some education and training in interpersonal facilitation knowledge and skills for the health care ethics consultant is not to argue that an ethics consultant should be "cross-trained as clinical psychologist."[29] Neither is it to argue that learning the tricks of the trade of facilitation is a substitute for the substantive knowledge of ethical problem-solving. It is to argue that would-be health care ethics consultants should be aware of any insufficiency as regards their knowledge of interpersonal facilitation and that they should take appropriate measures to remedy any such insufficiency, prior to engaging in this work. Failing in this regard, they will be unable to deploy their other expertise, to the detriment of those seeking assistance.

Conclusion

The health care ethics consultant's knowledge and skill in the area of interpersonal facilitation (negotiation, adjudication, mediation) can be understood, in certain respects, to be more essential than the consultant's other forms of knowledge and abilities. Lacking this ability, for example, the health care ethics consultant will be unable to initiate or conclude suitable discussion or consultation concerning resolution of the presenting ethical dilemma. In their discussion of case consultation, West and Gibson provide the following case:

> Assume that the family of a comatose patient takes the position that life-support treatment should be terminated immediately. The doctor, on the other hand, may insist that more information is needed before making that kind of decision. Delving beneath the outcome sought by the family, the [consultant] might find such a position based on the anger, frustration, and pain that comes from dealing with what the family perceives as an uncaring, unresponsive institution rather than from a basic wish to let the patient die. The family's underlying interest may be in ending a situation that is unbearable for the family members. On the other hand, in advocating delay, the physician may be looking not so much for more information but rather for emotional and legal support. The doctor's underlying interest may be to avoid legal liability or to be faithful to a moral commitment to sustain a patient's life. In this situation, where the patient's family members feel that their concerns have not been heard and responded to by the physicians and hospital staff, the [consultant as mediator] might help the family and hospital design a method for future communication that must be implemented before the family is ready to consider the more basic treatment issues.[30]

The ethics consultant who attempts to assist here using intellectual facilitation only may well fail to be of real assistance. Clearly, consideration of alternatives for action, value assumptions, consequences of actions if taken, and ethical rationale will all be essential in helping those who "own" this dilemma. At the same time, none of this activity will be possible, much less fruitful, unless the ethics consultant also has some skill in the area of interpersonal facilitation. Thus, how does one communicate with angry, frustrated people? How does one empathize with those who feel pain and require emotional support, while also moving the ongoing discussion of difficulty to conclusion? What interpersonal techniques will facilitate the intellectual facilitator's work? How does one learn and improve on the use of these techniques?

Again, interpersonal facilitation serves as the recognized medium within which the ethics consultant's knowledges and skills are assembled, and as the vehicle that carries the consultant's other forms of knowledge and skills (in clarification, analysis, and assessment, and as regards terminology, resources, and so on) for deployment to best advantage. Thus, while not sufficient in its own right, ability in interpersonal facilitation can nonetheless be seen to carry more than its own weight in the consultation process.

In this view, preliminary examination of the apparent current understanding of the nature and role of interpersonal facilitation (negotiation, adjudication, mediation) in the work of ethics consultation provides grounds for certain uneasiness. How competent are individual health care ethics consultants in this matter?

As grounds for such uneasiness, consider first that facilitation is often cited and well-recognized as a knowledge and skill essential to health care ethics consultation. Already noted is the fact that such facilitation has both an intellectual and interpersonal aspect, that the intellectual aspect is well-described and pursued in both the literature and curriculum offerings, but the same level of interest and information does not apply to interpersonal facilitation. The presence (or absence) of discussion of a topic in the literature is one of the criteria currently used to judge the concern of a profession regarding a particular subject.

Applying this criterion to the interpersonal aspect of facilitation, there is evident inconsistency between the claim that facilitation is essential to ethics consultation and discussion of it in the literature (or any preparation for it in practice). Such inconsistency should be eliminated, lest the credibility of the profession be called into question.

Second, and following from the first concern, there is a certain uneasiness regarding the adequacy of service to clients. Those who seek out health care ethics consultants regarding mediation, for example, are seeking assistance in matters of grave concern to them. These individuals are vulnerable in terms of their need; they are trusting in terms of their expectations. If the ethics consultant is not truly competent to assist in such interpersonal facilitation, but nevertheless holds himself or herself out to be a competent facilitator, that consultant is guilty of bad faith. Any such implicit deception of individuals may harm them, even as it denigrates the practice of health care ethics consultation.

Third, there is concern for interprofessional collegiality. Many of those who act as consultants in the health care setting have achieved that status by way of a curriculum that included formal study and training in interpersonal facilitation. Graduates of recognized health science curricula and other health care team members, for example, social workers, pastoral care workers, and psychologists, have recognized competence in this area by way of a credentialing process that accredits their practice. If health care ethics consultants choose not to follow the credentialing path in interpersonal facilitation while claiming professional status in their consultative work, they should at least be aware that their level of professionalism is not equivalent to that of their colleagues in the health care setting.

To remedy the uneasiness regarding the competence of health care ethics consultants, those who hold themselves out to be health care ethics consultants must recognize that theirs should be a dual expertise. They require knowledge of theory and practice in both interpersonal and intellectual facilitation. It follows that although due attention must be given to study and training in intellectual facilitation, such study and training would be incomplete without due attention to the theoretical and practical aspects of interpersonal facilitation, and to the integration of these two pursuits. Stated differently, given professional claims and client needs regarding ethics consultation in general, and the role of interpersonal facilitation within the consultation process, on any apparent nonresearch in this area, any apparent nonattention to its study, any apparent lack of interest in the development of specific interpersonal facilitation skills, whether on the part of the individual (would-be) health care ethics consultant or the profession of health care ethics consultation more generally, can only be judged to be inconsistent, not to say unethical, behavior.

Notes and References

[1]Jacqueline J. Glover, David T. Ozar, and David C. Thomasma, "Teaching Ethics on Rounds: The Ethicist as Teacher, Consultant, and Decision maker," *Theoretical Medicine* 7 (1986): 19; Terrence F. Ackerman, "Moral Problems, Moral Inquiry, and Consultation in Clinical Ethics," *Clinical Ethics: Theory and Practice*, eds. Barry Hoffmaster, Benjamin Freedman, and Gwen Fraser (Clifton, NJ: Humana, 1989) 153; Françoise Baylis, A Profile of the Health Care Ethics Consultant, in this vol.

[2]Terrence F. Ackerman, "The Role of an Ethicist in Health Care," *Health Care Ethics: A Guide for Decision makers*, eds. Gary R. Anderson and Valerie A. Glesnes-Anderson (Rockville, MD: Aspen, 1987) 313,314.

[3]Glover 19.

[4]Jonathan D. Moreno, "Ethics Consultation as Moral Engagement," *Bioethics* 5.1 (1991): 55.

[5]John La Puma and David L. Schiedermayer, "Ethics Consultation: Skills, Roles, and Training," *Annals of Internal Medicine* 114.2 (1991): 155–160.

[6]The reference to those who "own" the moral problem is borrowed from F. Baylis, "Ethics Consultation: The Hospital for Sick Children Initiative" *HEC Forum* 3;5 (1991): 289. This notion may be fleshed out with reference to Christine Mitchell's response to the question "who should decide about the goals of care?" Her response includes: those who bear the burden of both care and conscience; those with special knowledge (technical knowledge and experiential knowledge); and those health professionals with the most continuous, committed, and trusting relationship. Christine Mitchell, "Care of Severely Impaired Infant Raises Ethical Issues," *The American Nurse* 16.3 (1984): 9.

[7]For a comparison of the various activities, *see* L. N. Rangarajan, *The Limitation of Conflict: A Theory of Bargaining and Negotiation* (London: Croom, 1985) 258,259; more generally, *see* Paul Wehr, *Conflict Regulation* (Boulder, CO: Westview, 1979).

[8]Cf. Blair H. Sheppard, Kathryn Blumenfeld-Jones, and Jonelle Roth, "Informal Thirdpartyship: Studies of Everyday Conflict Intervention," *Mediation Research: The Process and Effectiveness of Third-Party Intervention*, eds. Kenneth Kressel, Dean G. Pruitt and Associates. (San Francisco, CA: Jossey, 1989) 166–189.

[9]Cf. Steven J. Brams, *Negotiation Games: Applying Game Theory to Bargaining and Arbitration* (New York: Routledge, 1990); Howard Raiffa, *The Art and Science of Negotiation* (Cambridge: Harvard University Press, 1982); H. Peyton Young, *Negotiation Analysis* (Ann Arbor: University of Michigan Press 1992).

[10]Cf Jay Folberg and Alison Taylor, *Mediation: A Comprehensive Guide to Resolving Conflicts without Litigation* (San Francisco, CA: Jossey, 1984); Sarah Childs Grebe, Karen Irvin, and Michael Lang, "A Model for Ethical Decision making in Mediation," *Mediation Quarterly* 7 (1989): 133-48; Janice A. Roehl and Roger F. Cook, "Mediation in Interpersonal Disputes: Effectiveness and Limitations," *Mediation Research: The Process and Effectiveness of Third-Party Intervention*, eds. Kenneth Kressel, Dean G. Pruitt and Associates. (San Francisco, CA: Jossey, 1989), 31–52.

[11]Mary Beth West and Joan McIver Gibson, "Facilitating Medical Ethics Case Review: What Ethics Committees Can Learn from Mediation and Facilitation Techniques," *Cambridge Quarterly of Healthcare Ethics* 1 (1992): 63–74.

[12]West 64.

[13]West 71; cf William A. Donohue, "Communicative Competence in Mediators," *Mediation Research: The Process and Effectiveness of Third-Party Intervention*, eds. Kenneth Kressel, Dean G. Pruitt and Associates. (San Francisco, CA: Jossey, 1989) 322–343.

[14]Baylis, A Profile of the Health Care Ethics Consultation, p. 40.

[15]Folberg 35.

[16]Kent E. Menzel, "Judging the Fairness of Mediation: A Critical Framework," *Mediation Quarterly* 9 (1991): 3–20; cf Folberg 245–250, regarding fairness in the context of mediation in divorce and family disputes.

[17]West 66,67.

[18]West 71; Bernard Mayer, "The Dynamics of Power in Mediation and Negotiation," *Mediation Quarterly* 16 (1987): 78.

[19]Cf Folberg 249.

[20]West 69.

[21]Bethany June Spielman, "A Mediation Model of Clinical Medical Ethics," (unpublished).

[22]With reference to ethics consultation, cf West 70; Joel Frader, "Politics and Interpersonal Aspects of Ethics Consultation," *Theoretical Medicine* 13 (1992): 31–44; with reference to mediation more generally, cf Folberg 140.

[23]Folberg 263 ff.

[24]Raiffa 232.

[25]Folberg 232–243.

[26]Folberg 236 ff.

[27]Folberg 240.

[28]Folberg 237.

[29]Folberg 241.

[30]West 72.

Feeder Disciplines

The Education and Training of Health Care Ethics Consultants

Michael Burgess, Eugene Bereza, Bridget Campion, Jocelyn Downie, Janet Storch, and George Webster

Preamble

Traditionally, health care ethics consultants have come to the practice of ethics consultation from a variety of academic backgrounds. For example, there are those who pursued ethics through the study of philosophy and theology and those who entered the field through the study of medicine. Most of these ethics consultants have formal training in only one discipline, having enriched their disciplinary training by attending seminars in ethics, conducting sabbatical studies in ethics, or collaborating "on-the-job" with colleagues from other disciplines. Some of these consultants have interdisciplinary training (i.e., training in a program that integrates several disciplines, as in some bioethics programs), and a select few have multidisciplinary training (i.e., disciplinary training in two or more fields, as with a physician-philosopher).

In addition to those who have trained as physicians, philosophers, and/or theologians, there are anthropologists, lawyers, nurses, social workers, and sociologists of health care who are becoming more visible contributors to health care ethics activities. Along with graduates of programs specifically designed to produce bioethicists, they may also function as health care ethics consultants.

With the increasing diversity in background and training found among practicing health care ethics consultants, the creation of programs designed to produce bioethicists who may, in turn, become ethics consultants, and with the recent attempts to create a system of certification for the profession, the legitimacy of the feeder disciplines approach comes into question. Although the feeder disciplines may have served as the sole route to the profession, are they now a route of entry whose time has passed? In an era of specialization and certification, does this approach not seem somewhat haphazard? Should feeder disciplines be treated as a historically interesting concept, but decidedly an inferior method of producing qualified health care ethics consultants?

The underlying contention of this chapter is that the feeder disciplines approach is indeed a legitimate and sound means of developing common ground for health care ethics discourse without losing the field's multidisciplinary richness. At its simplest, the discussion of feeder disciplines is a road map to a common place from various points of departure—law, medicine, nursing, philosophy, and theology. Common language and opportunities to collaborate are necessary for members of the professions and the academic disciplines who focus their expertise on problems in health care ethics. This discussion of the feeder disciplines focuses on how each of the selected disciplines contributes to the common ground, and develops special strengths in some elements of the Profile.

The feeder disciplines selected for discussion reflect the disciplines that have been the major contributors of health care ethics consultants to date. There is no intent to exclude other disciplines. Other disciplines and different learning experiences may be adequate, or even excellent, for the practice of health care ethics consultation. For example, the increased attention to ethics within various disciplines may produce students of those disciplines who serve as health care ethics consultants. Also, life experience outside formal academic training may contribute to one's preparation as a health care ethics consultant, or supplement an academic program. These are highly individual points of entry, and no attempt is made here to characterize or evaluate them. Rather, the feeder disciplines approach is an open invitation to other disciplines to develop alternate routes or points of departure.

What follows then are individual accounts of how a student in law, medicine, nursing, philosophy, or theology can achieve the common ground necessary to health care ethics consultation. These accounts are not the result of empirical studies, but represent the perspective of a member of the particular discipline. These perspectives were shaped by several interdisciplinary discussions comparing disciplines and sharing perceptions about them.

Process

The authors of this chapter struggled to use the Profile to evaluate the strengths of each discipline relative to the Profile and to develop an understanding of what elements of the Profile required further training. First, each author had to designate the "basic" level of expertise or certification for their discipline. The second task was to evaluate the level of preparation for the practice of ethics consultation available within the discipline's usual educational program. Finally, assisted by interdisciplinary discussion, each author was to estimate further educational experiences necessary to achieve the elements of the Profile.

Members within each discipline had difficulty finding a noncontroversial degree or certification for the practice of health care ethics consultation within their discipline. The interdisciplinary discussions identified three possible criteria for determination of the "basic" degree or educational experience within a discipline. The first criterion was credibility of the individual within the discipline—for example, what education would be required of the philosophy student, for him/her to be considered a philosopher by those who are qualified philosophers? For some disciplines, however, it seemed that credibility within the discipline did not establish credibility as a health care ethics consultant. The second criterion for determination of the basic degree within a feeder discipline was based on an evaluation of the level of expertise required within the discipline specifically for credibility to perform the activities of health care ethics consultation—for example, what education would be required of the nursing student for him/her to be considered a qualified health care ethics consultant by those who are qualified nurses? A third criterion was credibility as a health care ethics consultant independent of disciplinary credibility. The last of the three criteria was rejected because the feeder discipline approach uses the Profile to describe expertise for health care ethics consultation, and assigns the determination of professional or academic credibility to the particular discipline. As appropriate, the other two criteria were used to determine the basic degree for health care ethics consultation for each of the feeder disciplines.

For law, the LLB and articling are the standard the profession accepts as qualifying one to practice law, and for the discussion that follows, this serves as the "basic" qualification. Nonetheless, master's level preparation is recommended as background for ethics consultation. In medicine and philosophy, disciplinary credibility historically has correlated with the disciplinary qualifications of health care ethics consultants. This allows for use of the MD with an internship or residency and the PhD in philosophy as "basic" degrees for each of these disciplines. For nursing, the second criterion was used to identify a master's degree in nursing as "basic" qualification. Although increasing numbers of practicing nurses are completing an undergraduate degree in nursing, the master's degree is considered necessary for credibility within nursing to engage in academic and leadership activities that correspond with the activities of a health care ethics consultant. Finally, with theology the second criterion for the identification of "basic" degree was used. In theology, there are professional and academic degrees reflecting different types of "practice." It seemed most appropriate, therefore, to allow an internal determination of credibility for the role of health care ethics consultant.

After identifying the basic degree required for each discipline for practice as a health care ethics consultant, each author then sought to provide an overview of the requisite disciplinary training. For the purposes of this discussion, the authors assumed that students interested in health care ethics consultation would actively pursue relevant educational experiences available within, or additional to, their disciplinary program. Disciplinary programs are conducted within universities and it seemed reasonable to assume that students would tailor their education so as to acquire the necessary knowledge and skills.

Furthermore, the activities of the health care ethics consultant also seem to require some character traits that differ from the socialization of many of the disciplines. Ethics consultants may require further training to develop some of these elements of the Profile, and sometimes to defeat problematic professional or academic socializations. For example, hard-nosed analytic philosophers or theologians must learn to be sensitive to emotional and practical dimensions of ethical problems. Clinicians who are accustomed to efficient decision making that resolves one problem and quickly moves to the next must develop the patience to understand the full complexities of ethical issues before drawing conclusions. All of the disciplines are accustomed to some degree of authority based on their professional credibility. Consultants from all feeder disciplines may need to overcome authoritarian impulses in order to develop the character trait of humility necessary for the ability to mediate.

This chapter is a collection of individual accounts of the strengths that each feeder discipline brings to the field of health care ethics, and how each assists in the professional formation of a health care ethics consultant. The accounts have been shaped through discussions among members of other disciplines who have practical experience with multidisciplinary ethics consultation. These perspectives likely tempered the claims that any discipline could make with respect to the achievement of the Profile.

As a whole, the chapter reflects the richness of the field now known as bioethics, and realistically depicts the work of the health care ethics consultant as a part of a larger set of professional activities and as broader than any single discipline's activities. A preponderant intuition among those who worked with this material was that the soul of health care ethics consultation and of bioethics was rooted in its multidisciplinarity. The feeder disciplines approach holds the promise of preserving this richness and of nurturing an increasing diversity of approaches.

Law

Training in law typically provides students with an understanding of the system of rules that are binding on communities (local, provincial/state, national, and international). Through legal training, students develop their analytical and critical thinking abilities, problem-solving and verbal skills, and the ability to craft arguments. Thus, the education of a lawyer provides a base on which to build toward the Profile of the health care ethics consultant.

At present, there is no real consensus within the legal profession with respect to the level of expertise required for credibility to perform the activities of health care ethics consultation. Therefore, the following discussion of law as a feeder discipline refers to the training necessary to be considered qualified to practice law by the profession. This discussion describes what law school (leading to an LLB or other first law degree such as the JD in the US) and admission to the bar can reasonably be expected to contribute toward fulfilling the Profile, and what additional training would be required in order to fulfil the Profile.

Training in Law

An applicant for admission to law school must have completed two or three years of university education. However, few students are admitted with only these minimum requirements. Most students have completed a four-year undergraduate degree, many have done graduate work, and some have had considerable nonlegal work experience.

In contrast to the other feeder disciplines, law schools do not generally recommend any particular prerequisite course of study. As a result, students come to law schools with such diverse backgrounds as commerce, dance, economics, engineering, history, music, philosophy, police work, politics, and theater.

Law schools tend to look for the following qualities in their applicants: high intelligence; sound judgment; the capacity and motivation for demanding intellectual effort; the capacity and motivation to engage in sophisticated legal reasoning; and an understanding of and sensitivity to human interaction. As evidence of these qualities, law schools tend to look to the following factors: academic achievement; Law School Admission Test (LSAT) score; and nonacademic achievement (e.g., student politics, music, and sports). Increasingly, law schools also consider disadvantages that students may have faced and overcome (e.g., physical, racial, and socio-economic circumstances).

It usually takes three years to complete law school. In the first year, students typically take core required courses (for example, contracts, criminal law, and torts). In second and third years, most or all courses are optional, but students usually take such courses as administrative law, business organizations, evidence, family law, labor law, and tax.

Students interested in health care ethics consultation can focus their attention on courses, activities, and independent research projects in health law and health care ethics to the extent that the flexibility of the program and the resources available at the university allow. Some schools offer courses of direct relevance—e.g., medical jurisprudence, reproductive health law, and law and psychiatry, whereas others offer courses of indirect relevance—e.g., international human rights, legal theory, and children and the law. Also, some schools offer opportunities for students to gain practical experience that can be transposed to health care ethics consultation, such as clinic work.

The requirements for admission to the bar vary tremendously. For example, some states only require that law school graduates pass the bar exams. In contrast, some provinces require that graduates complete a 17-month bar admission program (part course work and exams and part articling or clerking). There is little (if any) opportunity connected with admission to the bar for students to focus their attention on health law and health care ethics.

Development of Elements of the Profile

Like physicians and nurses, most lawyers complete their professional training without focusing on health care ethics. The remaining discussion applies to lawyers who have used available elective and free time to pursue such a focus.

Knowledge

Graduates of law school who have been called to the bar and who have focused on developing expertise for health care ethics consultation can be expected to have, at least at a low level, some of the knowledge requirements of the Profile. Knowledge of the legal system, relevant government regulations, policy statements, legislation, and legal cases (K9)[1] can be acquired through health law courses or independent research in law school, and through health law cases during articling (although the latter would be rare). Knowledge of facilitation techniques and underlying theory (K5) can be acquired through mediation and negotiation courses in law school and through working with clients and observing negotiation, mediation, and arbitration during articling.

Knowledge of medical terminology, the health care system's structures and decision making methods, and relevant professional guidelines and codes of ethics (part of K6) can be acquired through bioethics courses in the philosophy department and health law courses in the law faculty. Some (but not extensive) knowledge of the current bioethics literature and the "classics" in bioethics (K1) can be acquired through bioethics courses in the philosophy department.

Some (but not extensive) knowledge of at least one ethical theory/tradition/cultural belief system that is rich enough to allow one to develop a style of thought, habits of rigor, and judgment (K2), and some knowledge of a number of ethical theories/traditions/cultural belief systems that are most commonly held by those with whom one will be working (K7) can be acquired through ethics courses in the philosophy department and independent research courses in the law faculty. For example, a student might take courses in ethics, utilitarianism, or Jewish law.

Some knowledge of one's own biases/partiality (K4) can be acquired through courses, such as Feminism and the Law, or Race, Culture, and the Law. In fact, many law schools have recently introduced such courses in an effort to increase student awareness of biases and partiality.

Despite these educational opportunities, however, graduates of "basic" training in law require further training to meet the knowledge requirements of the Profile. Most important, further training in ethics, and health care ethics, health care and cultural differences is necessary.

Abilities

Graduates of law school who have been called to the bar and who have focused on developing expertise for health care ethics consultation can be expected to have, at least at a low level, some of the abilities listed in the Profile. The ability to facilitate (e.g., mediate, negotiate, and arbitrate), coupled with the ability to ascertain when one or more of these activities is appropriate (A6), can be developed through mediation, negotiation, and arbitration courses in law school and through working with clients and observing negotiation, mediation, and arbitration during articling.

The ability to acquire relevant information when gaps are revealed, coupled with the ability to know which questions to ask when attempting to fill in the gaps (A2), along with the ability to recognize and work within the limits of one's knowledge balanced with the ability to accept challenges (A7), are important for survival during law school and the bar admission process. Law school and articling experiences often hone these abilities. However, it is important to note that graduates of law as a feeder discipline will have these abilities only in the context of law. They will need further specialized training in order to be able to use them in the context of health care ethics consultation.

The ability to communicate effectively with health care workers, patients, their families, administrators, and social agencies (A5) can be acquired through legal training for effective communication with clients (for example, clinic work and articling). The ability to recognize one's own partiality (part of A8) can be acquired through such courses as

Feminism and the Law, and Race, Culture, and the Law, mentioned above.

The ability not to introduce personal beliefs in an inappropriate manner (part of A8), the ability to participate in group decision making when this may generate a conclusion with which one disagrees, and a willingness to tolerate such group decisions coupled with a strong sense of personal and professional integrity so that one may distinguish outcomes that one will not, on moral grounds, endorse/sanction from outcomes that one disagrees with, but will endorse/sanction (A9) can be developed through discussions of professional responsibility during law school and articling that require students to grapple with the issue of having to represent positions or participate in activities that they disagree with. For example, students might be required to deal with the issue of what strategies they would be willing to endorse/sanction in the defence of a man charged with violent sexual assault.

The ability to withstand the influence of public opinion and to question existing traditions, customs, and laws (A10) can be developed through courses that challenge the status quo (for example, a course in public interest advocacy or a course in critical legal theory). Again, despite these educational opportunities, graduates require further training in order to meet the abilities required by the Profile. Most important, further training in philosophy (specifically ethics) and exposure to the clinical context are necessary.

Character Traits

The law school admissions process, in principle, relates to some of the character traits in the Profile (for example, an understanding of and sensitivity to human interaction). In practice, however, academic performance often determines admissions. As such, too much should not be drawn from the mere fact of admission to a law program.

Also, completion of law school does not guarantee the development or existence of any of the character traits. What character traits particular students have will often depend more on the character traits of the students' teachers and peers than on the program itself (insofar as students often copy what they see). In addition, what character traits particular students have will often depend on what they went into law school with rather than what they acquired through the program. It is not at all clear that any of the character traits in the Profile are particularly fostered by law school. In fact, some have claimed that positive character traits are driven out of students by law school—for example, some have claimed that humility and integrity are driven out of law students by the elitism and competitiveness of many law schools.

The process of admission to the bar, in principle, also relates to some of the character traits in the Profile (for example, integrity). If someone is called to the bar, then he or she has been judged to have honesty/integrity and trustworthiness. In practice, however, there is no effective judgment system in place to ensure that all those admitted to the bar possess these character traits. As with admission to a law program, too much should not be drawn from the mere fact of admission to the bar.

High standards of character and conduct have long been, and continue to be, perceived as essential prerequisites for entry into the legal profession. As in medicine, ethical character and conduct were thought to come about as a result of mentorship. However, again like medicine, history has effectively demonstrated that faith in the mentorship model was not always warranted, and that unethical conduct at the individual and collective levels was not always prevented. Some attention is now being paid to the need for ethics education within legal training. Some law schools and bar admission programs already include a legal ethics education component. Other law schools and bar admission programs are considering introducing legal ethics education. However, legal ethics education has a long way to go before it can be relied on as a source for character traits for health care ethics consultants.

Summary

Although a law school and bar admission program shaped by a student who wishes to focus on health care ethics may give a lawyer some of the abilities, knowledge, and character traits listed in the Profile, no program will provide students with all of the elements of the Profile. Furthermore, no program will provide these elements at a sufficiently high level to justify considering the lawyer qualified to act as an ethics consultant.

The results will vary according to how much time and effort the student devotes to the project of becoming qualified to act as a health care ethics consultant. The results will also vary according to how flexible the law school is with regard to a specially designed program, to whether the law school has (or has access to) strong courses in ethics, bioethics, and health law, and to whether the student must article or simply write bar exams.

Even with a specially tailored program, law school and bar admission program graduates will need additional training to develop the requisite knowledge and abilities. They will require philosophy training (specifically in ethics and bioethics) and clinical exposure in order to be qualified to act as health care ethics consultants. They will need to pursue supplementary education, including graduate courses in ethics and bioethics, and supervised work in a clinical setting (observing ward rounds, observing research ethics committees in operation, participation in in-hospital ethics education programs, doing consults under supervision of a practicing ethics consultant, doing rounds, reviewing charts, presenting at Grand Rounds, and working on policy development).

Medicine

"To cure sometimes, to relieve often, to comfort always." This 15-century folk saying may be the most famous and concise description of the essential nature of medicine and the role of physicians.[2] Others have attempted to provide more comprehensive descriptions.[3] Traditionally, medicine is perceived as both an art and a science. It is dedicated to the betterment of humankind, whether through prevention, treatment, or palliation of illness, disease, and suffering. The scope of modern medicine is broad, ranging from basic scientific research, clinical care of patients, training of new physicians, policy development in health care, and political activism.

Physicians face the constant challenge of integrating their scientific knowledge of medicine with their experiential knowledge of health care. The minimum requirement for such integration to practice medicine is internship or residency training. The same requirement could be reasonably expected of a physician contemplating the role of health care ethics consultant.

Training in Medicine

The following discussion describes how graduate-level training in medicine can meet some of the requirements of the Profile. There are several entry routes and types of training available for the practice of medicine. Invariably, the training is a long, involved process, accompanied by a sequence of evaluations intended to assess a candidate's suitability for progression through the various phases of medical training. A typical physician will have completed a three- or four-year undergraduate degree (usually in the sciences), a three- or four-year medical degree, a two- to five-year residency training program, and often further postresidency or fellowship training in a subspecialty area.

There is some variation in entry requirements for students applying to medical schools. Different schools place varying emphasis on academic grades, MCAT scores, specific science prerequisites, autobiographical letters and letters of intent, letters of recommendation, and interviews. A few schools consider academic performance exclusively and favor younger college students. Most, however, attempt to evaluate candidates on the basis of their intellectual abilities, academic record, personality, character traits, and overall suitability for the profession. There is a pervasive sense that most schools still emphasize the importance of academic performance in a traditional science undergraduate degree. Such an emphasis lends itself to a quantifiable and rigorous evaluation of one aspect of a candidate's suitability. There is far more variability in the methods and rigor with which a candidate's personality and character traits are evaluated.

Although innovative approaches to medical curricula appear to be gaining momentum, most programs are based on a progression from basic sciences to clinical sciences to clinical rotations. In medical school, courses in history-taking and physical examination increasingly emphasize the need to be sensitive to issues of patient autonomy, cultural diversity, gender, and so forth. Of primary importance, however, is the requirement that the student learn a systematic approach to problem-solving in the clinical context. This approach requires the development of many skills, including acquisition and evaluation of relevant information, consideration of options and their consequences, and formulation of a rational management plan. The method requires an efficient, critical, and comprehensive approach, as well as the dedicated and open-minded attitude associated with true scientific inquiry. In order to apply this method well, a student must also learn interpersonal communication skills, an appreciation for the concept of a therapeutic doctor–patient relationship, and the ability to function as a member of a team. This training is usually provided at the bedside by medical faculty who act as teachers, role models, and mentors. In the course of numerous clinical rotations in different clinical settings, students are typically exposed to a wide range of role models, but are left to determine which mentor or composite role model they choose to emulate. Also, some medical schools dedicate teaching time within their curricula to areas that relate more directly to the humanistic components outlined in the Profile. This might include lectures or courses in the history of medicine, the sociology of medicine, medical anthropology, health law, and medical ethics.[4]

Acceptance into residency training is typically based on a global assessment of a candidate's intellectual abilities, clinical skills, attitudes, character traits, and personality. Different residency programs are perceived to require or attract candidates with different personal profiles. Family medicine, pathology, pediatrics, psychiatry, and surgery, for example, are perceived to require different constellations of specific strengths and attributes. Thus, although only a small percentage of candidates are asked to withdraw from medical training early in their careers, many are counseled or screened for admission into particular programs based on such global assessments.

Residents acquire a wealth of experience through their intensive exposure to a wide range of patients and clinical settings. The hallmark of residency training, however, is increased clinical responsibility. As they continue to extend their knowledge and develop skills, residents become directly involved in critical decision making, often with significant consequences for their patients. The skill of developing a trusting doctor–patient relationship becomes indispensable at this stage. With this in mind, residents are routinely evaluated to assess their overall suitability for eventual medical practice.

The mentor model predominates in residency training. Recently, there have also been attempts to include formal teaching of health care ethics. The Royal College of Physicians and Surgeons of Canada has approved, in principle, the requirement for ethics education if residency and fellowship programs are to remain accredited.[5] The American Board of Internal Medicine has also addressed the issue of identifying and evaluating "humanistic qualities" in their residents.[6] The specifics of such education have not yet been uniformly mandated or implemented, and individual programs are assessing how best to meet such recommendations.

Development of Elements of the Profile

Knowledge

Most physicians have completed medical school and residency training without focusing on health care ethics. The remaining discussion applies to physicians who would have used available elective and free time to pursue such a focus.

As with other feeder disciplines, medical students interested in health care ethics consultation would presumably enroll in courses and pursue individual tutorials in this area. Medical faculties in North America are usually located in well-established universities where faculty expertise in health care ethics, health law, philosophy, and religious studies would be available. Nevertheless, medical students and residents have relatively few electives within their curricula to take advantage of such opportunities.

Physicians, by virtue of their medical training, can be expected to have an extensive knowledge and critical understanding of medicine and medical practice. This would include an appreciation of the concepts of "health," "illness," "clinical practice," and "medical research" (K3), knowledge of current and emerging problems, and organizational aspects of health care systems, as well as relevant professional and institutional guidelines, policies, and codes of ethics (K6). They would also have a good knowledge of available resources in the community (K11).

Physicians can also be expected to have a practical knowledge of various ethical, traditional, and cultural belief systems most commonly held by practitioners, patients, administrators, and social agencies (K7). They may have an intimate knowledge of the human dimensions of ethical problem-solving, especially an understanding of the emotional responses to health problems (K8), as well as knowledge of cultural differences relevant to beliefs about health care (K10).

Physicians will have acquired much of this working knowledge through their clinical training and practice. Physicians who focus on ethics during their training can supplement this practical knowledge with a more extensive, academic knowledge of these areas. Courses in psychology, sociology of medicine, and medical anthropology, as well as individual tutorials in health care ethics can be taken as prerequisite courses or as electives. Similarly, physicians can combine their limited practical knowledge of health law with courses in ethics and jurisprudence currently taught in many medical schools (K9).

Even with supplementary courses, however, physicians are unlikely to have extensive knowledge of the health care ethics literature (K1), extensive knowledge of facilitation theory, or practical knowledge of facilitation techniques (K5). They are also unlikely to have a critical understanding of the philosophy of science or extensive knowledge of even one ethical theory (K2). They are, however, likely to have a practical, working knowledge of the "medical model" of health care that may be rich enough to allow them to develop a style of thought, habits of rigor and judgment.

Physicians are often faced with a struggle to balance influences in their professional development that are potentially in conflict. They are trained to recognized their own limitations in terms of knowledge, skills, and attitudes, as well as the current limitations of medical science. The professional socialization process, however, also fosters confidence, an appreciation of the gravity of one's responsibilities, leadership skills, and a commitment to medical progress. It is a challenge to balance these elements in a way that facilitates an appropriate self-critical attitude. Given this challenge, physicians would probably benefit from an opportunity to study their own values and biases extensively (K4).

Abilities

The abilities listed in the Profile are not specific to a particular discipline or profession. Proficiency in these abilities is probably best appreciated on a continuum, and may be achieved through a combination of innate talents, academic study, and practical training.

Physicians are trained to solve problems in a systematic, comprehensive manner. This approach is rooted, however, primarily in scientific methodology and in the "medical model" of clinical management. Physicians may feel that any inherent gaps in this approach, especially with respect to humanistic factors, may be compensated for, at least partially, by innate abilities or through clinical experience.

Generally speaking, physicians are trained to acquire relevant information when gaps are revealed (A2), and to recognize and work within the limits of their knowledge, balanced with the ability to accept challenges (A7). They may also claim that the systematic thinking inherent in their medical problem-solving methodology can be partially transposed to ethical problem-solving (part of A3).

Physicians might also claim, on the basis of innate talent and clinical experience, some proficiency in all of the other abilities listed. Although most physicians are not specifically trained in ethics, they perceive medicine as an inherently ethical vocation. Ethical medical practice is not perceived to be contingent on formal, academic training in ethics. A good physician should be able to identify and sensitize others to ethical issues (A1), make and defend sound ethical judgments (part of A4), communicate effectively with others (A5), facilitate and participate in group problem-solving (A6, A9), recognize one's own partiality (A8), and withstand the influence of public opinion (A10).

Some physicians will claim further expertise in some of these abilities. Physicians who work in particular disciplines or in multidisciplinary teams (e.g., family medicine, geriatrics, palliative care, psychiatry) may be particularly skilled in interpersonal communication skills, sensitizing others to ethical issues, as well as participating in and facilitating group discussion. These abilities may be a function of a combination of factors, including training in a specific department, personality, attitude, work load, and other demands.

Although most physicians can claim some competence in most of these abilities, there are significant shortcomings that can be addressed through formal training in ethics. Ethical problem-solving requires knowledge and skills that are not inherent to the scientific method or the medical model, but are more properly the domain of philosophy and theology.

Without formal training in ethics, physicians are unlikely to be sophisticated in their abilities to identify all ethical issues (A1), analyze the meaning of philosophical concepts and principles, identify underlying assumptions in philosophical arguments (A3), or outline important values and provide the best arguments for and against various options (A4). Although physicians should be able to make sound ethical judgments with respect to many aspects of their clinical practice, they may have considerable difficulty in making such judgments with respect to highly contentious or developing areas that pose new challenges (A4).

Furthermore, the socialization process within the medical profession may actually undermine some of these abilities. Physicians are often perceived as being less than adequate in their overall sensitivity and interpersonal communication skills (A5). Physicians, even when working in multidisciplinary teams, are often perceived as having the ultimate legal and ethical responsibility for a patient's care. Together with the typical stressors of work load, time constraints, and leadership role, this responsibility may lead physicians to be less willing to elicit, appreciate, explore, and clarify the evolving viewpoints, beliefs, and values of others (A5). As a result, they may be less able to provide a recommendation "for consideration" without attempting to manipulate the decision making process (A4), or to facilitate, participate in, or tolerate group decision making (A6, A9).

Physicians who have focused on ethics in their training may have improved in some or all of these abilities, depending on the nature of their focus. Those who would consider working as health care ethics consultants would nevertheless benefit from more formal training in ethics, as well as interpersonal communication skills.

Character Traits

Traditionally, high personal standards of moral character and conduct were perceived as essential prerequisites for entry into the discipline. Ethical conduct was thus presumed as a minimal requirement, to be reinforced and refined primarily through the mentor model. History has effectively demonstrated that such presumptions were not always warranted, and that this approach has failed to prevent unethical behavior from occurring at the individual and collective levels. Recently, there has been a growing trend to provide more formal ethics training at the undergraduate and graduate levels, presumably as an attempt, in part, to redress the situation.

There is a great deal of confusion with respect to the standards of physicians' character traits. Some feel that physicians as a group are not as sensitive, ethical, or skilled in interpersonal communications and professional relations with patients as they are expected to be. Reports of patient dissatisfaction with physicians, of physicians' abuse of patients, and of unethical behavior in clinical practice and research contribute to this perception.[7]

Others perceive these examples as rare exceptions to a fundamentally ethical profession.[8] Traditionally, physicians are assumed to be professionals with high standards of moral conduct as well as character traits conducive to sensitive, ethical interaction with patients and colleagues. Professional codes of ethics dictate some of these standards, and physicians are expected to adhere to them throughout their lives. Deviation from these standards is subject to professional review and sanction. Furthermore, individual physicians and professional associations are playing a key leadership role in developing the area of health care ethics. Although there is some evidence to support both perceptions, there has been no comprehensive, systematic evaluation of the ethical standards of physicians or the medical profession.

Assessment for entry into medical school and residency training often includes instruments designed to evaluate these features. Such evaluation is rarely comprehensive, and is not universally or consistently applied. Entry into the medical profession does not guarantee high ethical standards or character traits, nor does the mentor model of training physicians claim to guarantee the achievement of desired character traits.

Summary

Medicine is a core discipline in health care. It represents a substantive base for the issues that are of concern to health care ethics consultants. Physicians with residency training who have focused on health care ethics have an intensive experiential, as well as academic knowledge of health care issues, and so are well placed to undertake training as health care ethics consultants. Further training in health care ethics, health law, and philosophy, as well as in interpersonal communication skills could contribute greatly in this regard.

Nursing

Nursing's primary focus is the human response to actual or potential health problems. The role of nursing involves attention to the client's/patient's response to health risks, illness, disease, and disability, as well as the interventions and outcomes relative to these conditions. Thus, the education and experience of diploma nurses provides a strong base of knowledge and experience in patient care, which is enhanced in baccalaureate nursing education. However, for credibility and competence in health care ethics consulting a master's degree in nursing (MN or MScN) is recommended.

Types of Nursing Education Programs

Although baccalaureate nursing education is becoming more common, the basic requirement for entry to practice in the profession continues to be the completion of a nursing diploma. Several types of nursing diploma and degree programs exist. Nursing diploma programs (normally two to three years in length) are still offered by some hospital schools, although the majority of diploma nursing education is offered through community colleges. Nursing baccalaureate degree programs are of two types: the post-diploma baccalaureate degree program, in which a hospital school diploma graduate or college diploma nursing graduate studies for two to three additional years at a university to obtain a bachelor's degree in nursing; and a four-year (generic) BN/BScN program, which a student may enter from high school. There are increasing numbers of master's and doctoral programs in nursing.

Students in diploma programs (whether hospital-based or college-based) are no longer required to provide service in exchange for their education. Also, over the years, a focus on "training" has given way to a focus on "education," removing the task orientation of many diploma programs and adding several liberal arts courses and basic sciences courses to the curriculum.

Baccalaureate nursing programs have traditionally included a wide array of course work in humanities and basic sciences. Humanities courses normally prescribed are sociology, psychology, and anthropology; science courses include biology, microbiology, anatomy, and physiology. Nursing courses include a substantial amount of clinical practice in acute care hospitals, long-term care agencies, public health agencies, and community settings (e.g., home care, well elder care). Since this practical experience is complemented by course work exploring human responses to actual or potential health problems, the BN/BScN graduates are well grounded in communication skills (part of A5), in problem-solving, and in recognizing their own limitations in nursing, and have a good beginning knowledge of cultural differences (K8, K10), health care systems (through lectures and field experiences) (K3), community resources, and professional guidelines and codes of ethics (K11, part of K6). However, they would not have the depth of understanding in these areas necessary for effective ethics consultation. In order to achieve that understanding, master's-preparation in nursing is desirable and considered necessary. It provides a broader base of knowledge and skill for practice, as well as credibility within the profession, with other professionals in health care, and with the community. Graduate education in nursing involves attention to enhanced critical thinking through research and exploration of concepts; a broadened understanding of philosophical underpinnings of nursing practice; an enhanced understanding of biological, psychological, social, and spiritual phenomena of patients and patient care; and in clinically based master's programs, a concentrated clinical practice at an advanced level.

Development of Elements of the Profile

Knowledge

Master's-prepared nurses possess a good understanding of the health care system, health care organizations, and institutional policies and decision making (part of K6). Nursing course work (and support courses) from baccalaureate to master's-level education provide nurses with a strong conceptual foundation regarding health, illness, and human responses to illness, including a good understanding of Western medicine (K3). As early as the first year of the baccalaureate program in nursing, students are introduced to sociological, psychological, and cultural dimensions of death, dying, and grief. Through patient care, case studies, clinical logs, and personal journaling with instructor feedback and discussion, baccalaureate students come to understand the patient as a person. Through graduate courses in nursing, social science, and/or humanities, master's graduates develop a good knowledge of cultural differences relevant to beliefs about health care, and a greater knowledge about their own biases and predispositions (K10, K4). They also have a good understanding of social and cultural circumstances affecting patients' and caregivers' emotional responses to a health problem (K8). Their nursing courses and their capacity to network (i.e., to utilize a range and variety of formal and informal contacts in the community) provide a good knowledge of available community resources and expertise (K11).

Knowledge about nursing's Code of Ethics, begun in BN/BScN (and even nursing diploma) programs, is enhanced by attention to various philosophies and belief systems in health agencies, and by greater understanding of other health professionals' Code of Ethics in master's program study. Most master's graduates would be familiar with the major ethical theories (K7), and those who are particularly interested in health care ethics consulting would take course electives from the Department of Philosophy on moral philosophy or bioethics. They would have reasonable knowledge of one ethical theory or tradition. However, one could not assume that they would have extensive knowledge of ethical theory or of the bioethics literature (K2, K1).

Knowledge of the health law system and relevant legal cases would be imcomplete (K9). It would be gained throughout the nursing curriculum in legal aspects of such matters as charting, resuscitation, advance directives, use of restraints, and consent to treatment. Policy statements of professional organizations, particularly those relating to provincial or state nursing associations or national nurses' associations, would be generally known by master's graduates, as would knowledge of key legislation (e.g., professional legislation).

In summary with respect to knowledge, the graduate degree in nursing would be the recommended entry for nurses into health ethics consulting. Master's-prepared nurses would have a good knowledge in many aspects of the Profile, but would require enhanced knowledge in bioethics literature, particularly the classics (K1), knowledge and critical understanding of at least one ethical theory (K2), knowledge of facilitation techniques (K5), and knowledge of the legal system (K9).

Abilities

Master's-prepared graduates in nursing are normally able to identify ethical issues and to sensitize others to these issues (A1). They have a relatively good ability (acquired through a communications laboratory or clinical practice) to acquire relevant information from clients and from other health professionals to fill gaps in their knowledge and understanding (A2). Given the emphasis in diploma and baccalaureate programs on the nursing process (essentially a problem-solving process), master's-prepared nurses would possess good skills in bringing systematic thinking to ethical problem-solving by using their professional methodology (A3).

Communication skills (A5) receive a strong focus in nursing education from diploma through to doctoral programs. Although nurses would not have depth of understanding in mediation, negotiation, or arbitration, nurses often tend to be mediators on nursing units or on health care teams, drawing on formal education in conflict resolution, and assertiveness training (A6).

Master's graduates in nursing can generally be expected to be clear about their limits, but willing to accept challenges. Thus, in preparing for or serving as a health care ethics consultant, they presumably would be aware of their limits (A7). They are adept at participation in group decision making (A9), and are relatively strong in recognizing and clarifying their own values and beliefs, and in respecting the beliefs and value commitments of others (A8).

In summary, master's-prepared nurses match the Profile in respect to abilities through their ability to identify issues, recognize limits, negotiate, participate in group decision making, and recognize their own partiality. They would require further development in knowing questions to ask to fill the gaps (A2), in analyzing concepts and principles, identifying assumptions and reframing ethical questions (A3), in making and defending sound ethical judgments, in outlining values and consequences of alternative actions (A4), in exploring and clarifying beliefs and values, representing those values and perceiving interests that influence behavior (A5), and in questioning existing traditions, customs and laws (A10).

Character Traits

Early discussion of nursing ethics revolved around the virtues that a good nurse should possess with the assumption that good character led to good practice. Only 10 years ago, nursing legislation in several jurisdictions still contained statements implying that a nurse must be of good character.

Although some undergraduate programs continue to interview applicants, seek references, and/or require them to write an essay about their reasons for choosing nursing, others are resorting to grade point averages as key entry criteria. However, given the extensive clinical supervision provided in undergraduate nursing programs, considerable screening for suitability for nursing practice does occur in clinical practice and, in cases where the student nurse is deemed unsafe, unsuitable, or lacking moral character to practice nursing, she or he is counseled to change to another academic program. Although such screening cannot ensure ideal character traits or high ethical standards for all students, at least some are thereby discouraged from entering the profession.

Since "to nurse" is "to nourish," traits of kindness, gentleness, sympathy, and patience are perceived to be critical traits for nursing, as is respect for others. However, power struggles in health care can engender impatience, as well as a certain degree of apathy and indifference.

Honesty and integrity are traits expected of nurses as are prudential judgment, humility, trustworthiness, and friendliness. Fortitude and courage are traits generally only sanctioned for nurses within the power relationships of health care if they do not result in any threats to prevailing authority (formal or informal) in the health care system. For example, nurses involved in whistle-blowing about incompetence or unethical practice of others often lose their jobs, their status, or their reputation. Since in cases such as these nurses pay a high price for ethical behavior, these character traits may need nurturing. In summary, many of the character traits of master's-prepared nurses would be congruent with those listed in the Profile for health care ethics consultants.

Summary

The desirable credential for nurses to feed into health care ethics consultation would be the master's degree in nursing. This degree provides not only a strong knowledge and skill basis for consultancy for many graduates, but also provides the nurse with internal and external credibility.

Master's-prepared nurses would possess a heightened sensitivity to their client's issues and concerns, and would be cognizant of cultural differences, and different values and beliefs. They also would possess knowledge about the community, the health care system, health organizations, and other social service agencies. Their strongest abilities would lie in communicating with patients and families, mediating, recognizing their own limits, and participating in group decision making as opportunities arise. Character traits would include many of those listed in the Profile.

In order to assume a role as health ethics consultants, master's-prepared nurses would need a period of study (one year or more) in bioethics in order to ensure they were well versed in the bioethics literature and well acquainted with ethical theories. They would also need a stronger grounding in health law. In addition, they would need to learn ethical reasoning, particularly with respect to identifying underlying assumptions, reframing ethical problems, examining courses of actions, representing various viewpoints, and making and defending sound ethical judgments. Such knowledge would best be gained by formal study that includes substantial supervised training designed to integrate academic learning with practice.

Philosophy

Training in philosophy typically provides a student with strong analytic abilities, particularly in the area of logic, and knowledge of historical and contemporary approaches to fundamental issues, such as the nature of knowledge. Graduate training in philosophy refines these skills, broadens the knowledge base, and deepens understanding in areas of specialization. A graduate student in philosophy will typically concentrate in a particular area of philosophy, such as metaphysics, epistemology, or moral theory.

This discussion of philosophy as a feeder discipline for the ethics consultant assumes that the graduate student in philosophy with an interest in health care ethics will take courses in theoretical ethics, applied ethics, and bioethics. It does not assume that the student will have an official specialization in bioethics.[9] The following discussion describes what a typical PhD program in philosophy contributes toward the requirements of the Profile and what additional training philosophy graduates will require. This latter set of educational experiences might be available during graduate training, but will likely require additional postgraduate experience.

Training in Philosophy

Most graduate programs in philosophy require a minimum undergraduate experience with philosophy, if not a major. The novice graduate student in philosophy will have had an introductory level course in philosophy, one logic course, and several courses in history of philosophy or in branches of philosophy (epistemology, metaphysics, ethics, aesthetics, philosophy of science). A graduate student has potential for an ability to systematically analyze issues, an aptitude for reading the philosophical literature, and an ability to challenge and to accept criticism.

The most common training in philosophy, in North America, is within the analytic tradition. Full-time graduate training in philosophy to the PhD level is typically a four- to five-year program, although students often require additional years. The program would typically include a few large lectures, many small seminars, considerable reading and writing, and one or two theses, depending on the requirements for the MA. Often, required courses, prerequisite or qualifying examinations, language requirements, and logic occupy most of the nonthesis writing time. Seminars and papers typically have a strong analytic focus. Philosophy students study argument form and validity in logic and all other courses. This develops a strong ability to identify issues, suppressed premises, relevant information, validity of arguments, and to relate historically important discussions to a broad range of philosophical issues.

Presumably, a graduate student interested in health care ethics consultation would focus considerable attention on courses in ethics (and in bioethics, if available), philosophy of science, and perhaps philosophy of law. Directed reading courses would also enable more detailed study of specific topics and the development of some research skills.

Development of Elements of the Profile

Knowledge

Graduates of PhD programs in philosophy who have focused on developing expertise for health care ethics consultation typically will have a considerable knowledge of the current bioethics literature (K1). This will vary among graduates, however, depending on such factors as available faculty, student determination, and other opportunities. Graduates will have extensive knowledge and critical understanding of at least one ethical theory and will have developed a rigorous approach to ethical issues. It is difficult to determine whether this knowledge will support a capacity for judgment (K2), although some knowledge of one's own biases or partiality should have developed (K4). Many graduate seminars in ethics now utilize health care examples, developing some understanding of ethical frameworks and social dimensions of the various actors in the health care system (K7, K8), including professional guidelines and codes (part of K6). The same courses may also provide background in some cultural beliefs about health care (K10). Students who take courses in philosophy of law may develop knowledge of the legal reasoning, how law relates to morality, and principles of judicial reasoning. These courses are less likely to produce students who are knowledgeable in the specifics of the particular legal system, health law, and government regulations (K9).

A person who has prepared for health care ethics consultation via the PhD in philosophy is likely to be dependent on colleagues for other knowledge elements of the Profile. Without further training, a philosopher will require assistance from a health care ethics consultant or a clinically trained colleague for an extensive knowledge and critical understanding of "health," "illness," and "clinical practice" (K3).[10] Knowledge of medical terminology, common and emerging problems, the health care system, decision making, policies, and professional ethos (part of K6) will also require further training and collegial support. Knowledge of the legal system and relevant health law and regulations are likely to require further training or consultation in particular instances (K9), as will knowledge of the rich social and cultural circumstances shaping emotional and behavioral responses (K8).[11] Knowledge of available resources will require constant updating as new situations arise and services are discontinued or initiated (K11). Philosophers are unlikely to have theoretical knowledge of facilitation (e.g., mediation, negotiation, and arbitration) techniques and theory (K5), although systematic analysis and articulation of arguments in support of different positions are philosophical skills that may combine with certain personality types to provide some informal skills in this area (A6).

Abilities

A philosopher's strongest abilities will likely be the identification of ethical issues, the ability to sensitize others to these issues (A1), and the ability to bring systematic analysis to ethical problem-solving (A3).[12] Furthermore, philosophers' broad training in various moral theories and the controversial nature of all moral theories will enable them to defend others' ethical judgments and to present recommendations for consideration without investing too much personal interest in one position (A4, A8). Philosophy training will develop the ability to use the relevant information to identify the logical alternatives, and to provide the strongest arguments for and against each alternative (A4). Philosophers ought to be able to recognize the limits of arguments and to meet a wide array of intellectual challenges (A7). Philosophers are trained to teach controversial theories and examples, and they are expected to recognize and moderate the introduction of their personal values (A8, A9). Their ability to withstand the influence of public opinion and to question existing norms is typically quite strong (A10).

Systematic analysis (A3) includes identifying morally relevant information. Clinically inexperienced philosophers will require assistance in identifying controversial clinical judgments, and in knowing what questions to ask to reveal or resolve controversies (A2). Although philosophers are often able to identify what information is important to a discussion (cf discussion of knowledge), they have a reputation of being unable to communicate sensitively and nontechnically, and of lacking perceptual skills regarding interests, relationships, and social context.[13] Students concentrating on health care ethics consultation may find that discussions with other disciplines, public speaking, and clinical training would enhance this ability (A5). Finally, the ethics consultant with a philosophical background may tend to fall into an arbitration model, and would likely benefit from experience in mediation and negotiation (A6).

Character Traits

Philosophy programs do not generally claim to be professional schools that evaluate the character of their graduates. That is not to say that philosophers do not care when students lack personal integrity, are dishonest, or lack other virtues. A PhD in philosophy is simply not a basis for character assessment. Sometimes philosophers are perceived as authoritarian because of their defense and critique of positions, or perhaps because they are sometimes closed to the views of others. With appropriate experiences and "modeling" by instructors, a critical philosophical understanding of ethics would lead to humility in application and serve to guard against moral authoritarianism.

Summary

A PhD in philosophy provides a solid foundation for a health care ethics consultant. Without further training, persons could be of considerable assistance as philosophical consultants. Additional opportunities to achieve the remaining elements of the Profile are necessary for independent functioning as a health care ethics consultant. The nature of most of the knowledge and ability elements of the Profile that a philosopher requires suggests the need for a "clinical" learning period. This might also provide the philosopher, colleagues, and supervisors with an opportunity to evaluate whether the individual had any deficiencies of character parallel to the assessment of professionals.

Theology

Theology provides a foundation or starting point for the study of bioethical problems. Bioethics is concerned with ethical questions in the life sciences, questions about our origin and creation as a species, our relationships with those in the healing professions who care for us when we are ill, and the principles that guide practitioners in their care of those who are dying. Theological reflection by a "community of faith" about the reality of God and the dignity of the human person expresses deeply held convictions concerning our humanity, the sacredness of life, and our destiny. This reflection on the part of a religious tradition or community of faith can provide a rich context for the analysis of bioethical questions.

In theology, there are both academic and professional degrees, and each of these degrees typically corresponds to a difference in professional practice. Those who complete academic degrees often teach in a university, school of theology, or seminary. By comparison, those with professional degrees are, more often than not, engaged in pastoral ministry or pastoral care (e.g., parishes, hospitals, prisons, community agencies). The following is a brief description of the graduate academic (PhD) and professional (DMin) degrees in theology, as relevant to the practice of health care ethics consultation.

Graduate Academic Training in Theology

Theology, concerned with the relationship between God and humankind in this world, has always counted moral theology as an essential component. It is a practical undertaking, dealing with human conduct in light of revelation and ultimately the achievement of salvation. Coupled with this is the existence of denominational health care hospitals and other institutions established by people as expressions of their faith commitment and operated according to the mores of their particular tradition. Given this interest in practical ethics and health care, it is not surprising that theology has been a traditional entry into the field of health care ethics consultation.

This section will discuss the merits of academic theology as a feeder discipline, describing the program generally, evaluating it in light of the Profile, and providing a brief summary. This will be a subjective view of the discipline and its potential as an entry into health care ethics consultation. Also, it should be noted that, although the theology under examination is Christian, this is not intended to deny the value of the rich and diverse theological traditions of other faiths.

Because undergraduate training in theology is rare, many students will begin formal study of the discipline at the master's level. There are several degrees available to the master's student: a master of arts in theology (MA), master of theological studies (MTS), and a master of theology (ThM), the latter of which usually requires that students have had previous training in theology. The duration of master's programs varies from one to four years, and most expose students to the breadth of theology through mandatory core courses in such areas as scripture, history, and systematic theology before allowing them to specialize. In most cases, ethics falls under the auspices of systematic theology (although some schools may list it separately as "Moral Theology," or include it in another category, such as "Religion and Society"), and is almost always considered a part of the core curriculum. Thus, theology students are usually exposed to the fundamentals of ethics as a matter of course, and those who wish to may choose it as an area of specialization, perhaps tailoring some of their course work and thesis (if required) around issues in health care ethics.

Students may continue their study of academic theology at the doctoral level, pursuing either a PhD in theology (PhD) or doctor of theology (ThD). Depending on the program and the school, students may be required to choose both major and minor areas of study, meet language requirements, and participate in graduate seminars. With the completion of their course work (a two-year residency is common), they will demonstrate their competency, often through written and/or oral comprehensive examinations, and write a thesis.

Because of the multifaceted nature of ethics, students specializing in the area will have a number of courses from which to choose, each with its particular focus—ethics taught from a feminist or liberationist perspective for instance, historical surveys, and courses that examine issues in bioethics. They may be exposed to ethics in non-Christian religious traditions; they may also take courses through other faculties and departments that may approach ethics from a secular perspective. In directed reading courses, they may concentrate on specific issues in health care ethics and study with moral theologians engaged in health care ethics consultation. At graduate seminars, doctoral candidates may participate in the lively exchange of ideas that follows the presentation of papers. Ultimately, the student interested in health care ethics will have the opportunity to research, write, and defend a thesis, and so make a contribution to the discipline.

Development of Elements of the Profile

Knowledge

Because the study of theology presupposes a faith basis, but often takes place in an interdenominational setting, introspection is an integral part of the undertaking. Some courses will go as far as to require students to write reflection papers. Thus, students are expected to confront and come to terms with their own biases (K4), no matter what their area of specialization. Because ethics is treated as an essential component of theology, students will have exposure to basic ethical theories and traditions. By specializing, students will have the opportunity to deepen their knowledge of ethics and think critically (K2). Should they wish to study health care ethics specifically, they may, through course work and directed reading, begin to amass a knowledge of the literature and foundational cases (K1). In the wider university setting, students may extend this knowledge by attending seminars in health care ethics held by other faculties (e.g., philosophy, medicine, and law); they may also take advantage of the growing holdings of bioethics literature being cultivated by other faculties. These resources can help students become more familiar with health care, its medical and legal aspects (K6, K9), and the sorts of belief systems that inform health care practice (K7). Because of the academic nature of the master's and doctoral programs, students could garner theoretical knowledge of such concepts as illness and health (K3), knowledge that would be enhanced by exposure to the clinical setting. However, without this exposure, students are unlikely to have a real understanding of what it is like to work through an ethical dilemma in practice (K8), a process that affects lives and evokes emotions in a way not often witnessed in graduate seminars.

Abilities

Students in the academic theology program have the potential to come away with strong foundations in the theory of ethics, being able to identify and work through ethical dilemmas creatively and with a grasp of the relevant data (A1, A2, A3). The practical nature of theological ethics means that the task of ethics is not simply to raise questions, but to help find answers, prescribing a way of life in this world (A4). As a consequence of the interdenominational setting in which much of theology is studied, students often come face to face with their own biases as they converse with people whose beliefs may be very different from their own, and this provides an opportunity to develop tolerance (A8). Also, like all graduate students, those studying theology must recognize their limits while daring to accept challenges (A7). However, it is only through actual experience that students will assimilate the skills needed to help people work through ethical dilemmas—abilities such as being able to communicate clearly with individuals and in groups, to build bonds of trust, and to help a group wrestle with a dilemma (A5, A6), all the while maintaining one's integrity in the process (A9). To hone these abilities, students must not simply read about the practice of health care ethics consultation; at the very least, they must be exposed to the practice of it, and at the time of this writing, opportunities for such exposure are very limited in an academic theology program.

Character Traits

Although some programs will admit students on the strength of personal essays and references as well as grades, usually it is academic performance that counts, and so there is little way of knowing which character traits students have on entry to the program and which they cultivate along the way. Furthermore, although theology (unlike other arts programs) accommodates and may even presuppose a faith commitment, it is difficult to predict how this will manifest itself. The sense of following a calling may evoke humility in one individual and rigid dogmatism in another. This sort of unpredictability, along with the program's emphasis solely on academic achievement, makes it difficult to assess the strengths of the PhD student in light of the Profile's character traits.

Summary

Students in a graduate academic theology program will have the benefit of rigorous academic training, which could give them the opportunity to deepen their understanding of ethics generally and health care ethics specifically. Because theology has traditionally served as a feeder discipline for health care ethics consultation, students may find themselves working with professors who have done foundational work in the field or who may be practicing health care ethics consultants. Theology students may also take advantage of the larger university resources to supplement their knowledge of health care ethics.

The major shortcomings of the program stem from lack of clinical exposure and experience. This is an academic program; students will be well grounded in theory and (probably) self-knowledge. They will not have much experience with working through a real dilemma in a group setting or in a clinical setting. They may be complete strangers to the culture of the hospital. As noted, this may be remedied by providing opportunities for students to have clinical exposure by working alongside a clinical ethicist. Furthermore, students of theology usually work in a milieu that presupposes and affirms their theistic beliefs, whereas much of health care ethics is practiced in a secular setting. It may be very difficult for them to reconcile such very different world views or to think of those people who live and practice outside of their traditions as anything other than "anonymous Christians"—which does little justice to the real beliefs and values held by others.

Finally, no matter what preparation students may have, it can be a shock to move from calm campus waters to the sometimes stormy seas of the clinic. However, with their solid background in theory and an interest in putting ethics into practice, students of academic theology are in an excellent position to make the transition.

Graduate Professional Training in Theology

The doctor of ministry degree offers advanced study for persons who are already engaged in various forms of ministry in the church or community, and who usually have a basic professional degree in theology (MDIV). The Doctor of Ministry (DMin) degree is granted by universities or theology schools accredited by the Association of Theological Schools (ATS), and most doctor of ministry programs accept candidates from different religious traditions.

An important aspect of the doctor of ministry degree is the emphasis placed both on academic work and practical experience in ministry. Applicants must have a number of years of experience in the practice of ministry before being admitted to the program, and students must continue work in their ministry during their course of study. The expectation that students continue with their particular ministry while pursuing academic work helps to integrate classroom learning with practice.

Development of Elements of the Profile

Knowledge

This degree does not offer specialized training in bioethics or clinical ethics. It would be important for those entering this program of study to have had formal concentration in ethics or bioethics in an earlier degree program. Students interested in health care ethics consultation would have to select specific topics in bioethics as part of their course work or thesis requirements.

Knowledge of current bioethics literature could vary considerably from graduate to graduate (K1). Students are free (and encouraged) to take advance degree courses in other faculties (e.g., law, philosophy, education, sociology). These may include seminar graduate courses or directed reading programs. This interdisciplinary exposure may assist students in appreciating the viewpoints and beliefs of others (K7). Those who work in or enter this program of study from a health care setting would have some knowledge of health care and medical terminology, relevant institutional policies, and may be familiar with professional guidelines (i.e., standards of practice/accreditation requirements and professional codes of ethics) (part of K6). Also, they may be expected to have some knowledge of cultural differences in the health care system (K10). Students whose practice of ministry is in the health care field or clinical setting can also attend rounds and participate in continuing education programs organized by the hospital community. This may help students learn more about belief systems or values commonly held by clients (practitioners, patients, families, and so forth) (K7), and available resources (e.g., community support systems) (K11). It is difficult to assess the extent to which graduates would cultivate or develop a style of thought or habits of rigor in judgment. This would be shaped by life experience, clinical practice, and previous study within a particular theological/religious tradition. Graduates who focus on bioethics or clinical ethics in this program would, in all likelihood, have extensive knowledge of at least one "tradition of thought" (K2). Furthermore, collaborative work throughout the program might aid in sensitizing students to their own biases (K4).

Graduates of the program who have limited experience in health care would have to spend time in the hospital or clinic. In terms of the Profile of the health care ethics consultant, this clinical experience is essential. In the clinical setting, students can acquire a basic working knowledge of hospitals and the health care system (K6). In some programs or centers, students participate in teaching rounds and consultations (under supervision and with permission), and come to appreciate the variety of disciplines engaged in patient care (K7). Students who participate in extended practice will also have a deeper understanding and appreciation of the nuances and complex nature of ethical problems that arise in this milieu (e.g., neonatology, critical care, chronic care, geriatrics). Direct clinical experience might also cultivate a healthy regard for ambiguity. The professional degrees in theology are designed to help graduates achieve a high level of competence in the practice of ministry. Graduates acquire specific skills for ministry, but may not have formally studied certain subject areas in philosophy (e.g., epistemology, logic). Students can choose these courses in their elective segments but this may not provide sufficient depth or grounding to meet the requirements of the Profile.

Abilities

The doctor of ministry degree integrates university-based courses, seminars, and research at the advanced degree level with ministerial practice. In the university setting, students participate in a collaborative learning group that encourages independent study and provides a unique opportunity for support and critical reflection (A5, A8). Also, a "ministry base group" of persons from the students' ministry setting is established, and regular collaboration with members of the group is encouraged. This peer learning process enhances interpersonal skills and strengthens some abilities outlined in the Profile. These include recognizing and working within the limits of one's knowledge (A7), recognizing one's own partiality, and not introducing personal views in an inappropriate manner (A8).

In addition to the collaborative learning group and the ministry base group (both of which are usually interdisciplinary, and may include men and women from other cultures and religious traditions), some programs require students to develop an individual learning plan in their first year. In terms of the Profile, this experience may help students develop certain abilities, including the ability to acquire relevant information when gaps are revealed (A2), and the ability to sensitize others to the beliefs and values of others (part of A5).

Many in the doctor of ministry program have extensive pastoral experience. Some will have completed advance-level training and certification in clinical pastoral education with concentration in counseling (parts of A4, A6, A9), group dynamics (parts of A5, A6), or special problems in pastoral ministry (i.e., bereavement/care of the dying).

Character Traits

It is difficult to assess the extent to which formal programs assist in the development of "desired" character traits. Students admitted to the professional doctoral degree must have the support of their religious community or denomination and most candidates are interviewed by an admissions committee which considers such things as the candidates' pastoral experience and theology of ministry. The "weight" given to letters of support and preadmission interviews varies from university to university, and some professional degree programs may focus more on "character" in terms of admission requirements. Any character assessment at the time of admission, however, in no way guarantees that students who complete a course of study will exhibit the qualities highlighted in the Profile.

Pastoral experience might cultivate gentleness, tolerance, sympathy, empathy, and most importantly, humility. Depending on pastoral experience and area of ministry, some candidates may also demonstrate courage and fortitude in their pastoral work and study. The requirement that students work collaboratively (not in isolation) with peers and those in their "ministry base" may highlight the importance of character traits listed in the health care ethics consultant Profile. The cultivation of desired character traits may also be influenced by role models, mentorship, and supervised extended practice in the clinical setting.

Summary

A strength of the graduate professional degree in theology is the collaborative approach to learning and the integration of academic work (courses and thesis) with the practice of ministry. In many doctor or ministry programs, students work closely with peers from different religious denominations. This provides an excellent opportunity for students to learn more about other religious and faith communities. The extent to which students would be exposed to non-Christian faith traditions, such as Judaism, Hinduism, Islam, or Buddhism would vary from program to program.

Graduates with a doctor of ministry degree would have some of the requisite knowledge and abilities of the Profile, but would very likely have to supplement this (if available) with graduate level courses in other disciplines, such as philosophy and law. Graduates would have extensive pastoral experience, but many would have to spend time in the clinical setting, observe and participate in rounds and consultations (under supervision), and also where possible, work with ethics committees and Research Ethics Boards.

Conclusion

The application of the Profile to each of the feeder disciplines has provided a sort of theoretical test of the Profile—the Profile having been used to identify the strengths of each of the feeder disciplines in the essentially multidisciplinary activity of health care ethics consultation, to indicate areas of weakness for each of the disciplines and show how these weaknesses might be compensated for by the strengths of other feeder disciplines.

As such, the foregoing application of the Profile to the selected feeder disciplines is an example of how individual students, various disciplinary programs, and specialized bioethics programs can apply the Profile to their circumstances. This use of the Profile and the feeder disciplines discussions avoids authoritarian proscription of qualified persons from practice and provides a means of developing the shared expertise necessary for competent health care ethics consultation while preserving the rich multidisciplinary nature of health care ethics.

Notes and References

[1]Letters and numbers in parenthesis refer to the Profile. "K" refers to the numbered item under "Knowledge," and "A" to "Abilities."

[2]Maurice B. Strauss, *Familiar Medical Quotations* (Boston: Little, 1968) 410.

[3]Edmund D. Pellegrino and David C. Thomasma, *A Philosophical Basis of Medical Practice* (New York: Oxford University Press, 1981) 58–81.

[4]Françoise Baylis and Jocelyn Downie, *Undergraduate Medical Ethics Education: A Survey of Canadian Medical Schools* (London: Westminster Institute for Ethics and Human Values, 1990).

[5]Royal College of Physicians and Surgeons (Canada), Office of Training and Evaluation, *Newsletter* 1.2 (1991).

[6]The American Board of Internal Medicine, Subcommittee on Humanistic Qualities, *Guide to Awareness and Evaluation of Humanistic Qualities in the Internist, 1991–1995* (Portland, OR: ABIM, 1991).

[7]E.g., The College of Physicians and Surgeons of Ontario, Task Force on Sexual Abuse of Patients, *The Final Report of the Task Force on Sexual Abuse of Patients* (Toronto: CPS(O), 1991).

[8]"Doctors Have Highest Ethical Standards, Respondents in Gallup Poll Say," *Canadian Medical Association Journal* 147 (1991): 1041.

[9]Although there exist programs in health care ethics within many philosophy departments, they are not the focus of this discussion.

[10]David C. Thomasma, "Why Philosophers Should Offer Ethics Consultations," *Theoretical Medicine* 12 (1991): 129–140.

[11]Barry Hoffmaster, "Can Ethnography Save the Life of Medical Ethics?" *Social Science and Medicine* 35 (1992): 1421–1431.

[12]Peter A. Singer, "Moral Experts," *Analysis* 32.4 (1972): 116,117.

[13]Hoffmaster.

From Avocation to Vocation

Working Conditions for Clinical Health Care Ethics Consultants

Benjamin Freedman

Other health care ethics consultants may have shared this experience. When I began my current clinical appointment, no formal orientation took place (no informal one either, as far as I can recall). Nobody sat me down to tell me what I was supposed to be doing there; if the question was raised at all, it was in the form of options or possibilities, joined by saying, "You're really the expert; you'll be teaching us." Nobody told me for whom I was working or to whom I was to report. (Some time after I started, somebody told me I worked for him. I told him—no hard feelings—he was mistaken. A year or so later, I decided he was mostly right and, from time to time, have occasion to reconsider yet again . . .) There were plans to have an Ethics Committee on which I would serve (or with which I would work—not settled), but there were no terms of reference established, a task that had awaited my arrival. There had been apparently a few preliminary discussions between my clinical and academic institutions about the relative proportions of my time that should be spent at each institution. These desultory negotiations had not resulted in agreement, and so the two decided to leave this up to me as well. I had been assigned neither office space nor secretarial assistance (not surprising, perhaps, since nobody knew what I was going to do in one or with the other). I was, however, given a choice of either a beeper or an answering machine, in order that I not miss any calls.

With a lot of patience, friendliness, and good will on the part of senior hospital staff, and the expected stubbornness on mine, things have so far worked out. It could have been otherwise, and on a couple of occasions, it looked as though it was.

The following comments represent a framework for discussing working conditions and expectations on behalf of persons employed as health care ethics consultants. The need for a discussion of reasonable, efficient, and justifiable working conditions is evident. The potential benefits may accrue to ethics consultants, those employing them, and those working with them. Working conditions openly agreed on in advance further everyone's interests (provided Byzantine and clandestine political machinations—themselves unethical—are ruled out as solutions).

Some literature has recently begun to appear on what institutions may expect of persons employed to do ethics consultation, but there remains very little if any discussion of the obverse: What might an ethics consultant expect of the employing institution? When unions or professional associations negotiate on behalf of their members, although most demands concern salary and perquisites, some demands delineate the minimal requirements needed to do a decent job well. Which demands would be appropriate on behalf of health care ethics consultants?

Ethical Exceptionalism: Two Straw Men, Easily Knocked Down

The language above, and its underlying concepts, are jarring. Ethics—a "job"? "Working conditions" for ethics consultants? Can these categories be meaningfully and appropriately applied to health care ethics consultation?

A case may be made for treating ethics consultation as *sui generis* —as exceptional to ordinary understandings of employment. For the first generation of ethics consultants, ethics consultation began as an adjunct to their primary duties of teaching, clinical practice, or research—almost a hobby. From their point of view, consultation provided the professional satisfaction of providing case material and an opportunity for field demonstrations for their real work, as well as the personal satisfaction of helping professionals and patients. To the doubly optional character of an ethics consultation—as regards calling for a consultation and following any of the ensuing recommendations—a third option was added: The consultant could respond to a request or not—indeed, could choose, without making any fuss or changing jobs, to discontinue his or her availability for consultations. This tradition persists for many of the most senior figures in bioethics, and within this tradition and its literature, it is puzzling to speak of working conditions.

Consider what may be the most crucial question associated with conditions of employment: On what basis could an employee be fired? According to many accounts, most baldly expressed by Terrence Ackerman, a health care ethics consultant's task is to assist others—especially physicians—in making decisions. Along with this account goes the belief that there is no, and could not be any, independent measure of an effective consultation. Since the doctor is the decision maker, anything the doctor judges as assisting the decision did in fact assist it, and if that is right, then the measure of an ethicist's effectiveness is the satisfaction of the physician requesting the consultation. What if the physician is pleased with the consultant because he or she has provided an ethical *imprimatur* on behalf of the corrupt course of action that was planned? What if the physician is displeased with the consultation because its resolution, though beneficial for a patient, complicated his or her life? It matters not.

Of necessity, in this model, the health care ethics consultant should work at the pleasure of (some critical mass of) physician consultees in the institution. And how could it be otherwise? Given the fact that the decision to request a consultation is optional on the part of the physician, why would a person be paid who is not consulted, and why would a consultant want to work in such a situation?

The upshot for a health care ethics consultant under this description is that the concept of working conditions is finessed—is, indeed, incomprehensible. If an ethicist could be dismissed by those whom he or she is to serve, under any conditions they deem acceptable, then *a fortiori* any working conditions less serious than job security must be entirely subject to those employing the ethicist as well. The ethicist plays Dagwood to the physician's Mr. Dithers.

How did we in fact achieve this odd result? What is it about this model of ethics consultation that makes it so exceptional to the work-a-day world—that makes the very concept of working conditions so inconceivable? The answer is simple. Under this conception, the health care ethics consultant is a slave. By definition, a slave has no rights to given working conditions. Should the ethicist be treated well, this would be adventitious—the luck of serving under a benign master.

One may fairly wonder whether this is a desirable model of work for ethicists alone among the working world. One may doubt whether the jobs of an ethicist—consultation, education, policy formulation—can be done well, or done at all, without the right to given working conditions. If one does not doubt this, the remainder of this chapter will be of no interest.

Seemingly, a similar result may be reached from the opposite view: the ethicist as moral hero. Although many difficult ethical issues coming to consultation have no clear response, some do; sometimes, no reasonable person will disagree as to the impropriety of a suggested course of action. An ethical consultant may see the job as assisting in clarifying issues within the gray zone and always being open to the possibility that such clarification may help determine the ethical course of action. Our ethicist, subscribing to a prescriptivist understanding of morality, takes the next step: The moral syllogism prescribes (or prohibits) a particular action.

There may be disagreement as to how often such clear-cut ethical issues—real moments of truth—arise; but my own experience and that of my colleagues demonstrate that they do. Sometimes because of inertia, the "system" grinding mindlessly away, or prejudice—sometimes even because of muddled thinking!—the choice is clear, and the wrong choice will in fact result in a patient's death. The ethicist, working within the channels provided, whatever they may be, has failed to secure the ethical choice. Is he or she not morally compelled to go beyond those channels, even at the cost of his or her job?[1]

Part of the definition of some jobs is the required response in extreme circumstances, at a moment of truth. A soldier must be prepared to put his or her life at extreme risk at any time: that is part of the job. Similarly, my strong belief is that an ethical consultant must be prepared for the possibility of losing his or her job at any time over some action needed to protect the interests of patients.

An ethicist-hero who is prepared to sacrifice his or her job at any time may be thought to be above considerations of working conditions, as an ethicist-slave is beneath them. What may count as a matter of conscience is limited only by the imagination; in principle, any aspect of one's work might be of crucial moral importance. Within this view, therefore, for the stout, self-reliant, morally autonomous ethicist, working conditions may be beside the point.

That this result is the same as was reached for the slave is no coincidence. To be a hero is to be a slave to one's conscience, or, we might say, rights to working conditions are of no use to the slave and of no help to the hero. I do not write for adherents of such a view.

On the other hand, one may feel that there is a difference between being prepared to resign at any point and being prepared to resign at every point, between thinking that anything may be a matter of conscience and that everything is.

Working conditions are set (or ought to be set) so decent people—neither slaves nor heros—may reliably do their jobs in a decent fashion. HLA Hart has noted that law may only play a role in conditions of relative vulnerability and relative altruism; those who are invulnerable would never need law to order their conduct, those indefeasibly selfish would never allow it to operate. In similar fashion, median conditions need to be specified on behalf of working conditions, established as of right, on behalf of health care ethics consultants.

The Internal Morality of Bioethics

One philosophical tradition that may be used to ground working conditions for health care ethics consultants is purposive rational reconstruction, as developed by Lon L. Fuller.[2] The most familiar application of the tradition is his "internal morality of law."[3] Fuller was asking the question, "What characteristics must be possessed by law, defined as 'the governance of human conduct by means of rules,' to be successful?" His response came in the form of eight principles. Law must, for example, be promulgated: A secret law could not serve to influence the behavior of the unknowing. The legal system cannot in general rely on retroactive laws (although the occasional retroactive law might be needed to rectify some earlier legal blunder): One's conduct today cannot be influenced by tomorrow's pronouncement.

Fuller claimed that these were principles of morality, and not simply of efficiency, because the law as an institution (and profession) embodied an ethic, specific to itself, but recognized and valued by society as a whole. The eight principles do relate to efficiency: They are the necessary means to the end of a legal regime. That end, however, in Fuller's view, is not morally neutral. Principles of efficiency can be described for any human endeavor, for example, bank robbery, but they are not principles of morality unless the end of the endeavor is itself morally valuable.

This template of purposive rational reconstruction can be applied to a variety of other activities. Fuller himself applied it to electrical engineering in a little-known address.[4] I have earlier described a generalization of the method and presented some applications to medicine.[5] How might it be extended to cover health care ethics consultation as a profession?

The question cannot be answered at this time, and there is a reason why it cannot be answered. Fittingly enough for a Fullerian project, the problem arises from unclarity about purpose—a failure in the internal morality of describing internal moralities. Some examples will help: If the point of law was the establishment of social order and absolute conformity, a different set of principles of efficiency would result. (Terror might be a necessary means.) If the point of bankrobbing as a profession is the enrichment of its practitioners, an internal principle requires stealth; if the point instead were to serve as an object lesson to bystanders, the opposite is called for.

The internal morality of health care ethics consultation, as of any other profession, relies on a specification of what health care ethics consultation is for—what a health care ethics consultant is supposed to do. Is he or she supposed to ponder, pose, clarify, or resolve moral issues, and which ones—those arising within a legal context, micro issues, such as consent or medical choices, meso issues, like the institutional allocation of resources, or macro issues, like different aspects of health policy? Whose issues? Who are the clients—the employing institution, doctors, health care workers, patients, or families? What tasks are performed—teaching, consultation, policy formulation, or research review? Each permutation of these possibilities will result in a different job description, and each job description implies a distinct rationale.

The task of delineating a profession's internal morality relies on a prior specification of purpose. There is no canonical description of the purposes a health care ethics consultant should satisfy to be found within the literature. From amongst the range of uncertainties associated with defining ethical consultation, Mark Siegler suggests that the root and most serious issue may be:

> lack of agreement on the goals of ethics consultation. . . . The wide range of possible goals for ethics consultation highlights the point that one reason we do not yet know what skills are required of the consultant and that we lack usable evaluation data is disagreement on the fundamental issue of goals and desired outcomes.[6]

Also, there is no pattern of employer expectations that may be explored, as the beginning story illustrates. A theory of internal morality must begin with a given purpose, which is not to be found for health care ethics consultation.

(Such proposed purposes must themselves be the subjects of moral critique; the "given" need not be "taken," but the form such critiques must take drastically differs from that appropriate to an "internal morality" exploration: different argument forms, standards of evidence, categories of conclusions, and so forth.)

In the light of these unanswered questions, what can be done as a rational purposive reconstruction of health care ethics consultation? Formally, the task of describing an internal morality of health care ethics consultation (in the absence of a canonical description of the job) reduces to describing conditions of employment necessitated by the above characteristics. What employment contract clauses are needed to permit information gathering (and so on) in a context of task uncertainty consistent with an employee's moral integrity? The following are some discussion points. As a start, I suggest we focus on three characteristics that are likely to appear in any reasonable description of a health care ethics consultant's job: communication, uncertainty, and moral integrity.

Communication

Notoriously, those jobs that we call "professional" emphasize information and communication (comprising information gathering, processing, and transmission) as central tasks. There seems nothing beyond these tasks essentially tied to health care ethics consultation.

Communication entails input and output: "hearing" and "speaking." Principles for each could be suggested. On "hearing," for example, a health care ethics consultant needs to have access to the information sources he or she believes to be relevant to the task. An ethicist asked to consult on a case, but not given access to the chart, or unable to speak with the patient, family members, or allied health professionals, and so forth, as the case may be, is placed in an impossible position. Such an *a priori* restriction, even if these sources of information in fact have nothing to provide, undermines confidence in the consultation. Discussion may need to ensue on which sources of information must be accessible, and whether there is a canonical list that may be provided (I think there might be intertheory agreement on which sources are potentially relevant, although for different reasons), and whether there are side constraints on such access that are justifiable.

So much for "hearing." What of "speaking"? One example: Before agreeing to work as consultant ethicist and writer for the Canadian Nurses Association, when that body was preparing its revised code of ethics,[7] I required of the committee members that they would agree in the discussions ahead of us always to "hear me out," whether or not my suggestions would ultimately be adopted. The point was illustrated well in a lecture of Rabbi Dr. Moshe Tendler. A story is told in the Talmud of Rabbi Ami's first class on his arrival in Israel.[8] Before responding (correctly) to a question posed, Rabbi Ami, about to give his inaugural speech, prayed that his words be accepted. He gave his answer, and the entire class laughed at him. Rabbi Moshe Sofer, a 19th-century scholar, commenting on this strange episode, asked why he was punished given that he had provided the right answer, and replied that Rabbi Ami was being punished for having sinned in his prayer: Rather than pray that his words be accepted, he should have prayed that his words be heard.

So it is in bioethics. For most of the interesting questions, of course, reasonable people disagree, and there can be no expectation on the part of the ethicist that his or her opinion, however reasonable, be accepted. However, it would be futile to work under circumstances where your opinion will not even be given a fair hearing.

Again, this seems straightforward and intertheoretically justifiable, but the details need to be worked out: Who has to listen, and at what length? Side constraints of confidentiality need to be specified, and the possibility of strategic side constraints explored, for example, blocking discussion of a case *sub judice* under the hospital attorney's legal advice.

These tasks will need to be taken seriously, for in the absence of a rational, well worked out system of communication on behalf of the health care ethics consultation process, we may anticipate that institutional pressures will come to bear with unfortunate results. Kenneth Simpson, in his paper, "The Development of a Clinical Ethics Consultation Service in a Community Hospital,"[9] illustrates the arrangements on communication that seem to result when instead of a rational approach, the dominant considerations are power, turf, and inoffensiveness.

He writes of his practice on "hearing," "Only the attending physician can request an ethics consultation, and access to the patient or members of the patient's family depends on the attending physician's consent." Two statements seem to be made: that a case can only come to the health care ethics consultant via a request by the patient's attending physician, and that once the referral has been made, that same referring physician's consent is needed to approach patient and family.

The first statement goes beyond the scope of this chapter. I cannot however let it pass without remark. Many of my colleagues have agreed to work under this condition: Formal ethics consults can only be requested by a physician (attending or house staff). As far as I can tell, the main effect of this restriction is to generate many quotation marks: When a request for discussion comes from someone other than the attending physician, the ethicist will call it a "consultation" instead of a consultation, or will say it was an "informal" rather than "formal" consultation.

Setting aside the fact that some of the most important ethical issues concern the propriety of the attending physician's own decisions and conduct, the presumptions underlying this condition seem to me wildly anachronistic. Although occasionally an ethics consult is born of an individual doctor's *crise de conscience*—for example, "Should I refer this patient who is requesting a late abortion for reasons of undesired sex to a facility that performs such?"—this is the exception rather than the rule. Usually, ethics consultations concern the ongoing treatment of a patient—treatment provided or at least influenced by treating physicians, consulting physicians, nurses, technicians, social workers, administrators, chaplains, and many others. If the input of these persons to a patient's care is important, it is beyond my understanding how their recognition of ethical dilemmas can be disregarded by administrative *fiat*. Please note that we have not yet even considered whether a patient's family member—or, save the mark, a patient—might wish an ethics consultation to discuss a treatment plan! Furthermore, of course, any of these can have their own *crise de conscience*. Thankfully, at my own institution (the Jewish General Hospital, Montréal, Canada), any member of the hospital family—including all levels of staff, patients, and significant others—may approach the ethicist or ethics committee to request a consultation; judgment is then exercised concerning whether, and in what manner, to grant that request.

The second fact—that the referring physician's consent is needed to approach the patient and family—is clearly unfortunate (if true), and would represent an intolerable restriction on professional practice. Bismarck once instructed his physician to examine and treat him without bothering him with any questions. The doctor responded that in that case Bismarck needed a veterinarian: Vets are used to treating patients who do not talk; doctors do not feel the same way. I'm not sure what to call a person who would do a consultation working from the written record and the word of an attending physician, without speaking to a nurse, a patient, a relative—a detective, perhaps, or a research librarian—but certainly not a health care ethics consultant!

Simpson's arrangements for "speaking" seem equally discomfiting, although for different reasons:

> Copies of the consultation report are distributed to the requesting physician, consultants, residents, and medical students involved in the care of the patient. In an effort to develop a working group that would retrospectively review and critique my consultations, I chose also to provide copies of every consultation report to the legal counsel, the medical director, and the chairman of the department of

> medicine. The hospital administration decided that a formal consultation report would not become part of the medical record unless specifically requested by the attending physician and first cleared by the legal counsel.[10]

Commentaries accompanying Simpson's piece[11] express dismay at the role assigned there to legal counsel. (That role is even more puzzling when we note that Simpson has been instructed "to refrain from legal commentary of judicial case citations"[12] in the consults he performs.) These commentators, however, are wrong to single that provision out. The entire arrangement described by Simpson—with all the many included in distribution, and the vital parties excluded, as described above—seems more designed to mollify power brokers within the institution than to further humane patient care rationally.

Uncertainty of Task and Authority

In the tradition of, "If you get a lemon, make lemonade," we should take as one starting point for discussion the fact of serious uncertainty surrounding the job of health care ethics consultant, and discuss what bounds need to be placed on uncertainty, as well as which conditions make that uncertainty tolerable.

It is tempting to say simply that the following are necessary: a clear and doable job description; clear lines of responsibility and authority; and no authority without corresponding power. These seem to be minimal management requirements for employer-employee relations, but the real world of work does not always respect dictates of rationality, and ethical consultants would not be alone in facing uncertainties about the source and scope of their authority. Perhaps more importantly, setting these matters in stone in advance fails to allow for learning, adjustments to local needs and expectations that evolve over time.

Consider one such question: To whom does the ethics consultant report? Ethics committees themselves have taken a variety of routes: the facility's executive director (or CEO); the hospital's board of directors or board of governors; the medical council or executive committee; and so on. The choice is commonly determined by the history of the ethics committee's development, but the choice has both philosophical implications, and practical (as well as political) ramifications. Are ethical dilemmas understood to arise as from within medical treatment or within the broader context of the institution's responsibility to persons within its census (to ask the same question in the cynical mode: is ethics to be a tool of the medical staff or of the administration)? Will the ethics committee report to the same body that determines its budget? Will oversight of the committee be done on broad measures of policy, or be subject to very close and detailed scrutiny? The choice of reporting lines is broader still for the ethics consultant; in addition to the above, further possibilities include the medical director or director of professional services, or the chief of the department of medicine (or of whichever department lodges the consultant's appointment), making an analysis of the implications of a choice, correspondingly more complex.

At the outset of an appointment, there is likely to be some office or individual that has taken the initiative in seeking an ethicist, in freeing up funds, in conducting the search, and so forth. A natural consequence is for that office or individual to then take on the role of "mentoring" the ethicist, and in the absence of any other candidates, the ethicist will see his or her line of reporting there as well. Even if that managerial structure was appropriate at the outset, though, any number of changes—in the mission of the facility, its power structure, and so on—may render it superannuated. (Also, the ethicist is likely to be the last to know that this change has occurred.)

A better approach might be to specify minimally necessary arrangements that need to be worked out in advance, and a framework for reconsideration and renegotiation to mutual advantage after a specified period. Under the first rubric, I am thinking of such homely arrangements as that the ethicist must be told in advance when he or she is going too far. Probably several have stories about the negotiation (or failure to negotiate) of uncertainty over power, responsibility, and authority in the context of cases: I know I have.

Individual Moral Integrity

Ironically, given the job under discussion, this is likely to be the most controversial characteristic of work as an ethicist, and the theoretical problems that it raises are formidable: What counts as integrity? Integrity to what? Who decides? Is the characteristic unique or more pronounced for this job than others? On a practical level, however, the demand for integrity seems to me more certain. In terms of the employer's own perception of the job, I think this characteristic ranks high.

This final *necessitatum* devolves at a minimum into the ability to do your job without compromising conscience. But we need not be satisfied with that minimum, instead seizing the opportunity to understand health care ethics consultation as a forum for expressing moral commitments, as other professions may provide a forum for expressing intellectual, aesthetic, utilitarian, or other commitments. Again, I suspect my own experience is usual: choosing bioethics at a time of rising social demands that the academy be socially and politically relevant, under the spur of a need to fulfill personal moral commitments and needs without abandoning academic interests and predilections.

Some of the elements that would go into this freedom of conscience are alluded to in my own "Bringing Codes to Newcastle: Ethics for Clinical Ethicists."[13] I describe there a story about how I came into conflict with a prospective employer on this account. In brief, during the interview for the position, I was asked about my response in the event of learning of some seriously unethical practice in the hospital. I suggested several in-house, ameliorative attempts to address the issue and was informed that they have been rejected. I responded, as I think any decent person should do, that I would take the necessary steps, including whistle-blowing, to stop the practice. I was on those grounds rejected for the job. It still puzzles me how an institution could think of imposing job restrictions on the exercise of moral conscience on the part of an ethicist. Does the principle of publicity make sense here—i.e., that the index of ethical suspicion is high if the actor would not want his or her actions to be publicly disclosed? Would that prominent medical center care to explain in public why they did not hire me?

As indicated, though, it is ironic that the idea that an ethicist must take a stand in the face of serious wrongs or injustices is far from accepted. The idea—which I had thought self-evident—that an ethicist should pursue moral commitments, may be rejected indirectly or directly. It is indirectly rejected when only temporizing approaches to moral imperatives are discussed. A good example of one who indirectly rejects the notion that an ethicist has moral commitments is Edmund Howe.[14] He writes of:

> whether the ethics consultant should assert a definitive position and insist that others adopt his position or remain more neutral, primarily clarifying alternatives. The former approach is more likely to protect an individual patient's immediate interests in the short run, but it may alienate the physician and deprive the ethics consultant of opportunities to help a greater number of patients in the long run.[15]

Howe's claim seems unexceptionable as stated, yet disquieting nonetheless. Certainly, prudence and indeed the deep commitment to morality accompanying it require that one exercise judgment in taking a moral stance. On the other hand, however, when does that judgment tip the balance in favor of ethical activism? Howe never tells us, even when speaking of clear-cut wrongs:

> the marked moral superiority of one approach and general consensus that this superiority exists are minimal (if insufficient) criteria for an ethics consultant to assert a patient's interests even over and against the interests of the referring physician.[16]

In fact, he maintains this stance against the obvious objection that an ethicist undercuts his or her own position by failing to advocate ethics personally:

> As a practical matter, the ethics consultant may lose credibility if he or she does not assert a strong position. . . . On the other hand, an effective ethics consultant should stress the existence and rationale of the ethical consensus rather than his or her own personal view, though both are in agreement.[17]

Howe concedes at the close of his paper:

> In cases in which patients are more severely harmed, such as when they are deprived of life-sustaining means or when they are kept alive against their will, the ethics consultant will probably be compelled to be an advocate for the patients' interests.[18]

Such cases are by no means rare: They are the meat and drink of ethics consultation. Yet, in 10 pages of discussion of "An Ethics Consultant's Responsibilities," this is all the direction one receives. Howe's comment begs the question: What "compels" a consultant to advocate a patient's interests? The answer is of course: the *raison d'être* of the profession.

James Drane takes the next step—to explicit rejection of the ethicist ever taking an autonomous ethical stance:

> Dictatorial and self-righteous behavior is inappropriate in any profession. No one profession has a corner on eternal ethical truths. If the physician is the professional who is primarily responsible for the patient, a medical ethicist with insight provided by

> the literary disciplines can offer help but cannot take over decision making or presume to tell the physician what to do . . . there will certainly be situations in which the ethicist advocates one opinion over another, but he or she should only perform this service when asked and should do so without appearing to take charge of decision making or assuming an air of infallibility.[19]

For Drane, then, it is never the case that an ethicist should attempt to intervene on his own hook, say, to save a patient's life. Drane has the courage of his misguided convictions. Fully six of his seven recommendations to those interviewing applicants to work in ethics consultation make the same point. The hospital administrator should avoid applicants who are:

1. Too theoretical . . . The best course of action in a particular clinical setting may be more accurately dictated by an experienced doctor's discretion than by a deontological or utilitarian theory of right and wrong.
2. Dogmatic. A clinical setting is a place of compromise . . .
3. Obsessive/compulsive. The perfectly right thing can seldom be done . . .
4. Too directive . . . The ethicist . . . cannot dictate the right course of action . . .
5. Too aggressive.
6. Too paternalistic. Good clinical ethicists do not insist on what they think is best for others . . . Doctors and nurses . . . must help people decide what is best for them.[20]

To say the least, then, there is no agreement among writers to the proposition that health care ethics consultants have a right, still less a duty, to speak from their own consciences and to act on that basis. Were this view to prevail, my task of describing working conditions for ethicists would be simplified. Since in my view persons working under this constraint defraud themselves, their employers, and the patients by claiming to act as ethicists, such persons have no right to work—hence, no right to working conditions. Since this is a problem of truth in advertising, the possibility remains open that other jobs, truthfully described, need discussion. What kinds of jobs? "Private confessor to attending physicians," for example; "junior assistant risk manager, cross-appointed to public relations." Needless to say, I have no interest or experience in these roles.

What is the real problem? Most will agree, after all, that institutions have no right to expect any employee to compromise his or her conscience. Ethicists are not, however, in the position of "any employee," most obviously, because the embroilment within morally sensitive questions is for them the rule rather than the exception. As alluded to above, in my experience, another factor looms large. By engaging a health care ethics consultant, an institution is making a new commitment to humane and just care. This is more than a perception shared within the hospital community; it is a perception fostered by advertising of the position, and by announcements of its having been filled, in public relations releases and hospital newsletters. Absent the intention to protect that ethics consultant's freedom of conscience, this is all a lie.

I think, however, that most institutions do understand the engaging of an ethics consultant to be an act of institutional self-reform and rededication. The problem is that this is too easily lost sight of once the honeymoon is over. The employer's intention to protect the ethics consultant from intrainstitutional pressures and commitment to maintaining the job security of one who acts conscientiously in the course of employment should be made concrete in writing from the beginning, as an explicit condition of employment.

How such a clause should be phrased is admittedly a difficult problem. Some may fear that an ethicist will be more adept than most at exploiting such a clause to cover deficiencies in other areas of performance; I do not disagree. Although I cannot resolve the specifics here, we may have made some progress if the principle has been accepted. As always, in designing conditions of employment, as was said above, we must keep in mind the average employee. A hero needs no protection of conscience, and a knave should not benefit from it.

One last word on this for those already employed without this protection. We know from the most extreme chapter in moral turpitude ever written by humanity that people confronted with the demand to act immorally are very prone to accede from fear of punishment, even when punishment in fact occurs rarely or never. German physicians who conscientiously refused to comply with directives to forward the names of their "mentally defective" patients for "euthanasia" did so largely with impunity. The massive conscientious objection of Dutch physicians to compliance with their Nazi occupiers brought no serious reprisal in the end. German soldiers who conscientiously refused to participate in Aktionen against Jewish civilian populations commonly had commanders who respected their wishes. In our own context, I have yet to hear of an ethicist who actually lost a job[21] because of a conscientious stance taken. I am genuinely afraid to contemplate to what extent this is because ethicists have been self-censors of their consciences.

An Example and Conclusion

In addition to the above conditions, specifically necessary for an ethicist to do his job (whatever *that* might be!), there may be more general working conditions needed as well. To establish whether these are valid conditions, a real relationship between them and the tasks of the ethicist must be shown. I will give one example of how the argument might proceed with reference to research in ethics.

I should distinguish at the outset, as urged by other members of our Network, between two forms of research. It goes without saying that an ethics consultant cannot competently respond to a request for consultation or policy formulation without researching and synthesizing the existing relevant literature. I am speaking rather of the kind of research that may lead to publication in an academic or professional journal, research of a breadth and intensity not *per se* needed to respond to a service request. I should also note that I am speaking of a working condition, not necessarily a requirement. Some institutions with strong academic affiliation may require of their ethics consultant, as they require of their attending staff, continuing publication as a condition of employment, but what if they do not—if, for example, the employer is a community hospital? Should protected academic research time be a routine condition of employment in engaging a health care ethics consultant?

The question of research time is likely to come up with growing frequency, as clinical career tracks are developed in parallel to academic ones. Must an ethicist working in a hospital or other health care facility have substantial time for research, or should the job conditions be restricted to service aspects of work? The ethicist finds it, perhaps, nicer and more congenial to have research and writing as part of the job. Is this a frill—a "perk"—or can a good case be made that it is a necessity?

To answer the question, we must assume something substantive about the ethicist's tasks. I will assume that clinical consultation on ethical issues is at least among them. The question is, then: Does dedicated research time make a strong contribution to consultation?

One answer: Academic research, like Gaul, is made of three parts: mastery of the literature relevant to the topic of research; an original insight, comprising a synthesis of, extrapolation, or departure from, that literature; and preparation of that insight in a form acceptable for academic publication. Each of these tasks corresponds to work done in consultation. Some literature always exists relevant to each ethical consultation (however banal the issue might be), and it is part of the ethicist's job to canvass the literature for relevant information or direction. The literature, however, is never fully adequate to the job of consultation: The expressed views are all over the lot; the issues dealt with are not quite on point; and so forth. Thus, every consultation will rely on the synthetic and analytic skills involved in research at least as much as the literature/bibliographic work. Finally, the consultation, just as academic research, will need to be explained in a concise, coherent, balanced—and, perhaps, convincing—fashion.

Moreover, at the same time that research provides useful practice for health care ethics consultation, it provides an evaluative reality check as well. Persons working in ethics are at least as prone to eccentric enthusiasms as others, and bear the further danger of being persuasive enough to carry their institution along. The discipline of submitting material based on these ideas for journal publication can yield constructive criticism, resulting in improvements or chastening rethinking. Indeed, this reality check is likely to be most needed in those employment settings that would tend to look at research with the least favor: isolated institutions with few or no colleagues in ethics and at best peripheral academic affiliation.

To decide whether dedicated research time is a reasonable working condition, we needed to make a substantive assumption about tasks, *viz.*, that consultation is required. Consistent with what was said previously, the grounding for other working conditions requires a "thick," substantive specification of tasks, rather than the "thin" theory of ethical consultation relied on above.

These are still early days in the transformation of ethical consultation from avocation to vocation. Over time, consensus should build, allowing for the construction of a thick description of job requirements. We cannot decide what tools are needed until we know what job needs doing, but the attempt will never begin until we arrive at a realistic appreciation of the vulnerability of ethicists within the clinical context.

Notes and References

[1]"At some point, the duty to report wrongdoing may even cause the ethicist to risk his or her job." John A. Robertson, "Clinical Medical Ethics and the Law: The Rights and Duties of Ethics Consultants," *Ethics Consultation in Health Care*, eds. John C. Fletcher, Norman Quist, and Albert R. Jonsen (Ann Arbor, MI: Health Administration, 1989) 168. NB: Lots there on duties; nothing on rights I could make out.

[2]The tradition's roots certainly include Kant's efforts at "transcendental deduction" of the conditions allowing for perceived experience and Descartes' derivation of personal existence from the fact of posing the question.

[3]Lon L. Fuller, *The Morality of Law* (New Haven: Yale University Press, 1964).

[4]Reprinted in Benjamin Freedman and Bernard Baumrin, eds., *Moral Responsibility and the Professions* (New York: Haven, 1984) 79–84.

[5]Benjamin Freedman, "A Meta-Ethics for Professional Morality," *Ethics* 89 (1978): 1–19; *see* further Benjamin Freedman, "What's Really Special about Professional Ethics: Response to Michael Martin," *Ethics* 91 (1981): 626–630.

[6]Mark Siegler, "Defining the Goals of Ethics Consultations: A Necessary Step for Improving Quality," editorial, *Quality Review Bulletin* 18.1 (1992): 15,16.

[7]Canadian Nurses Association, *Code of Ethics for Nursing* (Ottawa: CNA, 1985). (The Code was revised once again in 1991; the latter version is reprinted in Françoise Baylis and Joycelyn Downie, *Codes of Ethics: Ethics Codes, Standards, and Guidelines for Professionals Working in a Health Care Setting in Canada* [Toronto: Department of Bioethics, The Hospital for Sick Children, 1992] 40–50.)

[8]TB Beitza 39.

[9]Kenneth Simpson, "The Development of a Clinical Ethics Consultation Service in a Community Hospital," *Journal of Clinical Ethics* 3 (1992): 124–130.

[10]Simpson 125.

[11] Daniel J. Anzia and John La Puma, "Cultivating Ethics Consultation," *Journal of Clinical Ethics* 3 (1992): 131-3; Henry Perkins, "Clinical Ethics Consultation: Reasons for Optimism, but Problems Exist," *Journal of Clinical Ethics* 3 (1992): 133–137.

[12]Simpson 126.

[13]Benjamin Freedman, "Bringing Codes to Newcastle: Ethics for Clinical Ethicists," *Clinical Ethics: Theory and Practice*, eds. Barry Hoffmaster, Benjamin Freedman, and Gwen Fraser (Clifton, NJ: Humana, 1989) 125–139.

[14]Edmund G. Howe, "When Physicians Impose Values on Patients: An Ethics Consultant's Responsibilities," *Ethics Consultation in Health Care*, eds. John C. Fletcher, Norman Quist, and Albert R. Jonsen (Ann Arbor, MI: Health Administration, 1989) 137–148.

[15]Howe 137.

[16]Howe 138.

[17]Howe 143.

[18]Howe 146.

[19]James F. Drane, "Hiring a Hospital Ethicist," *Ethics Consultation in Health Care*, eds. John C. Fletcher, Norman Quist, and Albert R. Jonsen (Ann Arbor, MI: Health Administration, 1989) 117–134.

[20]Drane 127,128.

[21]As opposed to a job opportunity; of my own experience above.

Liability of Health Care Ethics Consultants[1]

Larry Lowenstein and Jeanne DesBrisay

This chapter is a result of cooperation between the members of the SSHRC Network and the law firm of Osler, Hoskin & Harcourt.

Early in the project, members of the Network considered the role of the health care ethics consultant and the standards by which health care ethics consultants should be judged if they were to come before the courts. In particular, there was interest in exploring the standards a court would use in deciding whether a particular individual fell below an acceptable standard of practice and thus became potentially liable for the negative consequences of a consultation.

Osler, Hoskin & Harcourt generously agreed to provide the Network with a chapter on the potential for legal liability of health care ethics consultants. The authors of the chapter met twice with a member of the Network in order to establish the parameters of the chapter and to discuss the nature of health care ethics consultation. They then conducted the necessary legal research and wrote a first draft. This was critiqued by the members of the Network, and revisions were made by the authors. This process was repeated once more before the final draft was completed. This back-and-forth process was necessary to ensure that the chapter would reflect the reality of practice in North America.

This chapter provides an overview of the potential legal liability of health care ethics consultants in Canada. The authors outline the major areas of potential liability—battery and negligence. They explain the legal concepts and general legal rules of battery and negligence and apply the particular features of health care ethics consultation to these concepts and rules. They conclude that the chances of a court finding a health care ethics consultant liable for damages are fairly small.

This conclusion highlights the relevance of the Network project. As the reader will see, the standard by which health care ethics consultants would be judged if they were to come before the courts is at least in part dependent on what health care ethics consultants believe to be the standard by which they should be judged.

Introduction

The following is a discussion of the potential legal liability of health care ethics consultants that might arise in the course of their professional duties. This discussion has been commissioned by the Strategic Research Network on Health Care Ethics Consultation funded by the Social Sciences and Humanities Research Council of Canada.

At present, there is no authoritative profile for a health care ethics consultant. There is no accreditation system, and no professional standards. A health care ethics consultant can work as a volunteer, as an employee of an institution, or as an independent contractor. Also, the health care ethics consultant may be a consultant to a caregiver, to a patient and his/her family, or to an ethics committee at an institution.

There are three typical activities performed by health care ethicists: case consultation, policy formulation, and ethics education. Of these activities, this chapter discusses the potential liability of a health care ethicist with regard to case consultation.

Many health care ethics consultants see their function in case consultation as one that includes a "problem clarification" (analytical) role. In this model, it is usually a health professional who asks for the health care ethics consultant's assistance, although on rare occasions, the patient or patient's family or friends may directly ask for assistance. The health care ethics consultant aims to explore with the health care professionals and/or the patient and family the relevant values and moral principles, including an understanding of the benefits and harms of alternate courses of action.

Some health care ethics consultants go beyond clarification of the ethical issues and view their role as also being one of "problem resolution" (decision making). It is important to note, however, that the line between the problem clarification and problem resolution models is not clear. Problem clarification may have a significant impact on the patient's ultimate decision. For example, framing an ethical problem in a particular manner or introducing an option that had not previously been considered may have some of the elements of the problem resolution model.

The work of the health care ethics consultant can theoretically be seen as lying on a spectrum with the pure problem clarification model at one end and the pure problem resolution model at the other. The role of the health care ethics consultant in specific instances will typically fall somewhere between these two extremes.

No case in either Canada or the United States has yet examined the legal liability of ethics consultants. Although in the 1986 case of *Bouvia v. Glenchur*[2] ethics committee members were named as parties, there was no discussion as to their legal liability in the reported decision. In *Bouvia v. Glenchur,* the petitioner sought an injunction for the removal of a nasogastric tube that had been forcibly inserted and maintained, against her will and without her consent, by physicians who were attempting to keep her alive through involuntary forced feeding. The California Court of Appeals overturned the trial decision and granted the preliminary injunction to Bouvia. The tube was removed as a result of the decision. In oral argument, her counsel advised that it was his belief that if the petitioner succeeded in having the tube removed at the preliminary injunction stage, she would not pursue the lawsuit further. The court also indicated that the preliminary ruling would probably resolve this tragic case.

There is a difference of opinion in the literature as to the potential liability of health care ethicists. At least one author has claimed in print that health care ethics consultants would be immune from lawsuits and have no legal accountability.[3] John A. Robertson states that malpractice suits against health care ethicists are so unlikely as to appear fancified.[4] Others have suggested that professional liability insurance should be purchased even if the current risk of a successful malpractice suit is minimal.[5] Other commentators are even less optimistic. They suggest that the potential exists for individual ethicists to be legally liable when conducting clinical ethics consultations. Since other nonphysician professionals who provide counseling or consultation services, such as psychologists and ministers, have been named in law suits, there is no reason to think that health care ethics consultants would be shielded from similar suits.

There are two main areas of the law in which medical professionals have been found liable in the past. If the medical professional is found to have been negligent in the execution of his/her duties, he/she will be held accountable for the damages he/she causes. Also, if the medical professional undertakes a procedure on a patient who has not given his/her informed consent to the procedure, the medical professional has committed a battery and will be liable to compensate the patient for the resulting damage.

The courts have recently curtailed the applicability of the tort of battery in the medical context.[6] Although previously patients attempted to recover for injury suffered as a result of medical malpractice in both negligence and battery, it is now more likely for a medical professional to be held accountable in negligence. In the case of an ethics consultant, there is potential liability for counseling a medical professional to commit battery on a patient, and therefore, the tort of battery will be discussed briefly.

Battery

The tort of battery is committed by intentionally bringing about a harmful or offensive contact with the person of another.[7] In the medical liability context, any medical treatment, apart from treatment in emergency situations where consent is not possible, is a battery if it is administered without the consent of the patient or someone having authority to consent on the patient's behalf.

Battery is restricted to cases where the health professional performs a procedure totally different from the authorized procedure or where he/she performs a procedure without obtaining any consent at all.[8] Battery applies to cases where the treatment for which consent was given was misrepresented and a different treatment was carried out. The cause of action in battery exists regardless of whether the patient has suffered harm and even if he/she benefited from the unauthorized treatment.

If the health professional fails to disclose to the patient all the risks of the treatment such that the patient's consent cannot be considered "informed," this no longer constitutes a battery. Instead, the health professional is found liable in negligence for a breach of his/her duty of care. When a patient gives his/her consent without all the material information, this does not invalidate the consent giving rise to a cause of action in battery.[9] The essence of the patient's complaint in a battery suit is the unauthorized touching on which a battery action is based.[10]

An essential element of the tort of battery is physical contact. The plaintiff need not be conscious of the contact at the time it occurred; it is sufficient if he/she discovers it after the event. Although courts have held that there need not be any actual touching of the plaintiff in that contact with the plaintiff's clothing,[11] with an object carried by the plaintiff,[12] or with a horse ridden by the plaintiff[13] can constitute battery, these actions all involve physical contact with something attached to the plaintiff or in very close proximity to the plaintiff. It is essential that the defendant do some positive and affirmative act; mere inaction such as a passive obstruction of the plaintiff's passage does not constitute a battery.[14]

No person has been held liable in battery for counseling another to commit a battery. Giving advice without any positive action would not involve the requisite touching. In a situation where a health care ethics consultant advises a health professional to perform an act that is later held to be a battery, only the health professional or the person who actually performs the touching to which the plaintiff did not consent can be held liable in battery, and even this is becoming increasingly rare.

Negligence

In order for a malpractice action to be successful against a health practitioner, four conditions must exist:

1. The health practitioner must owe the plaintiff a duty of care.
2. The health practitioner must breach the standard of care required by law.

3. The plaintiff must suffer a loss or injury.
4. The health practitioner's conduct must have been the actual and legal cause of the plaintiff's injury.

These are the basic requirements of any negligence action, and they will undoubtedly be applied by the courts when examining the liability of a health care ethics consultant.[15]

Duty of Care

The starting point for the determination of whether one person owes a legal duty of care to another is Lord Atkin's "neighbor principle" as outlined in the case of *Donoghue v. Stevenson:*

> You must take reasonable care to avoid acts or omissions which you can reasonably foresee would be likely to injure your neighbour. Who, then, in law is my neighbour? The answer seems to be persons who are so closely and directly affected by my act that I ought reasonably to have them in contemplation as being so affected when I am directing my mind to the acts or omissions which are called in question.[16]

It is extremely rare for a plaintiff to fail to show the existence of a duty of care in medical negligence cases. A duty of care would also presumably be found between a health care ethics consultant and a patient. It is foreseeable that any recommendation or clarification the health care ethics consultant makes could affect the treatment the patient decides to pursue. Even if it is the doctor and not the patient who requests the consultation, and even if the health care ethics consultant never meets the patient, the relationship between the health care ethics consultant and the patient clearly falls within the neighbor principle. Certainly the health care ethics consultant should have had the patient, "reasonably . . . in contemplation as being affected when [he] was . . . directing [his] mind to the acts or omissions which are called in question."

The duty of care between a patient and a doctor exists independently of any contract between them. In the context of a doctor–patient relationship, the doctor assumes a duty to use diligence, care, knowledge, skill, and caution in providing medical care.[17] The ethics consultant, like any other person who undertakes to act, advise, or provide services that could affect the lives or well-being of others in significant ways could be held legally responsible for negligent conduct when acting in the consultant role.[18]

If the health care ethicist is consulted by a health professional, he/she owes a duty of care to the health professional to take reasonable care. It is clearly foreseeable that by acting negligently, the health care ethics consultant could make the health professional vulnerable to a lawsuit or to professional discipline. For example, the health care ethicist may advise a health professional to perform a medical procedure on a patient that is later found to be a battery.

The *Family Law Act* also establishes a duty of care between the tortfeasor and certain family members of the victim of the tortious conduct.[19] The damages recoverable under a Family Law Act claim are limited to expenses incurred as a result of the negligence and damages for loss of care, guidance, and companionship.

In the case of a health care ethics consultant, the duty of care would be found to exist at common law both to the doctor and to the patient. A doctor could name a health care ethics consultant in a lawsuit in which the doctor is being sued for negligence. The health care ethics consultant could also be brought into a lawsuit directly by the patient or the patient's family.

Standard of Care

Although establishing that a duty of care exists will be a relatively simple task for a plaintiff, establishing that a given health care ethicist breached the requisite standard of care will be a difficult hurdle to overcome. Medical jurisprudence requires proof that the health professional fell below the standard of care reasonably expected of him/her before the plaintiff can succeed in a negligence action.

The method of proving negligence is to provide experts who will testify that the health professional has fallen below the standard of care. This standard of care will vary according to the training of the health professional, the prevailing standards in the local community, and perhaps also according to the expectations of the patient. The expert must also be prepared to testify that the health professional has not acted according to approved practice and has not simply made an error in judgment.

General Standard of Care

In medical cases, the standard of care is generally determined by the conduct of a reasonable health practitioner in the same field. In the leading case of *Crits v. Sylvester*, the Ontario Court of Appeal states:

> Every medical practitioner must bring to his task a reasonable degree of skill and knowledge and must exercise a reasonable degree of care. He is bound to exercise that degree of care and skill which could reasonably be expected of a normal, prudent practitioner of the same experience and standing . . .
>
> I do not believe that the standard of care required of a medical practitioner has been more clearly or succinctly stated than by Lord Hewart, C. J. in *Rex v. Bateman* (1925), 41 T.L.R. 557 at 559: "If a person holds himself out as possessing such skill and knowledge by or on behalf of a patient, he owes a duty to the patient to use due caution in undertaking the treatment . . . The law requires a fair and reasonable standard of care and competence."[20]

In *MacDonald v. York County Hospital Corp.*, the Ontario High Court described the standard of care required of a medical practitioner clearly:

> The standard of proficiency required of a general medical practitioner is that of the average, competent medical practitioner and if that degree of care and skill is lacking, a general practitioner will be found liable and negligent. The general medical practitioner is not judged by the standard of either the least qualified practitioner or the most highly qualified, but rather by the standard of the ordinarily careful and competent medical practitioner. Consequently, it is no defense to a defendant practitioner to show that he acted as he thought correct in the circumstances and to the best of his skill and knowledge, if he has failed to measure up to the standard of the average practitioner.[21]

According to a survey conducted by the Strategic Research Network and presented at the 4th Annual Meeting of the Canadian Bioethics Society, there are 12 self-identified full-time health care ethicists in Canada.[22] Given this small number, it is difficult to point to an accepted average standard of competence. Since the medical ethics consultancy profession is in its infancy, a court will have difficulty in establishing the experience and skill of the "ordinary" and "average" health care ethics consultant. There does not appear to be an "average" health care ethics consultant at the moment.

Although it is not necessary for a health care ethicist to work at the same level as the most experienced health care ethicist in the field, it would be prudent for any health care ethicist to aspire to this level of competence. Robertson has outlined a "minimum" standard of care:

> At a minimum, the ethicist would have to have a reasonable level of competency in describing, clarifying, and facilitating ethical analysis of a case, and then not overstep the bounds of analysis and clarification into normative ethics or substantive decision making in its own right. In short, the duty is to be a good, competent, reasonably humble ethicist. The

> malpractice would be in performing poorly in that role (e.g., omitting or misinforming about a key ethical issue, so that a major aspect of the case was missed, or preempting the decision by telling the physician what to do).[23]

As is clear from this quotation, Robertson sees the role of a health care ethics consultant as being strictly one of problem clarification, not problem resolution. According to Robertson, as the health care ethics consultant approaches the problem resolution model by making substantive decisions, he/she is "overstepping the bounds."

The Network has prepared "A Profile of the Health Care Ethics Consultant," which outlines its view of the necessary knowledge, abilities, and virtues of ethics consultants. The Profile indicates that a health care ethicist plays an integral part in the ultimate resolution of ethical dilemmas by reframing the moral problem in a manner that may suggest a resolution or by providing a professional opinion. The health care ethicist's job is not limited simply to a problem clarification role.

The difficulty with establishing a professional standard of care for the health care ethics consultant is that the profession itself has not yet been established. There is no professional body regulating health care ethics consultants or holding them to a certain standard. In this situation, a court may well turn to literature on the profession in order to establish the appropriate standard of care. For example, Robertson's view may be seen as the appropriate standard simply because this is the view with which medical ethics consultants may be familiar. By the same token, the standard articulated in the Profile prepared by the Network might also be relied on by a court.

Standard of Care of Specialist

The standard of care expected of health professionals is directly related to their qualifications. A specialist is expected to possess and to exercise a higher degree of skill in his/her particular field than would be expected of a general practitioner in that field:

> The law differentiates between the standard of care expected and required of a general medical practitioner and that of a specialist. The standard of proficiency required of a general medical practitioner is that of an average competent medical practitioner, whereas the standard of proficiency required of a specialist or expert practitioner requires a standard of proficiency of an average specialist or expert in that field. Obviously, an expert practitioner is expected to possess and demonstrate a greater degree of skill in his particular field than is a general practitioner.[24]

Ellen Picard summarizes the types of factors that might establish specialty status for a doctor:

> Evidence of education (degrees, certificates and memberships, publications and privileges) and training (internship, residency, research and special study) provides formal and relatively objective criteria for establishing specialization's status. In general, the greater the education and training, the higher the standard expected. Evidence of extensive experience in a specialty will certainly raise the standard and may even be a substitute for some of the formal criteria just mentioned.[25]

This principle may give the court some guidance in establishing a standard of care for health care ethicists. Ethical dilemmas are not a new problem in health care. The average, general practitioner has attempted to deal with these issues in the past without any special training and without consultation with an ethics expert. If a health care ethics consultant is presumed to be a specialist in clarifying and resolving ethical dilemmas, then the competence of the general practitioner in dealing with such issues may be used by the court as guidance to determine the higher standard of care necessary for health care ethics consultants. If a court first determined the standard of care of health professionals in dealing with ethical issues, the health care ethics consultant would then be held to a higher standard of care since the health professional is not a specialist in ethics. If a health professional did claim to be a specialist in ethics, he/she would be evaluated according to the standards of the health care ethics specialist.

The only difficulty with this argument is that traditionally a specialist is presumed to have skills surpassing those of a general practitioner. Although health care ethicists must have a minimum level of medical knowledge, many will not have the same medical expertise as a general practitioner. The health care ethicist would only be held to the higher standard of care for the specialist when making ethical decisions, not medical decisions.

In medical malpractice cases, the court establishes the standard of care for the specialist at the appropriate level of competence for the average health professional in the specialty. Evidence of extensive experience may raise the expected standard of care; inexperience will not lower it. Once the court has established the standard of care expected of the average health care ethics consultant, evidence of higher education, training, or experience may raise the standard of care expected of the particular ethics consultant. An ethics consultant new to the profession, however, without extensive experience in his/her field, or an ethics consultant with only minimal formal education in ethics will be held to the same standard of care as the average ethics consultant.

Since health care ethicists come from a variety of disciplines, another issue arises: Will a court apply a different standard for a physician-ethicist than a philosopher-ethicist or a lawyer-ethicist? The standard applied will be that of an average health care ethics consultant. The feeder discipline of the health care ethics consultant should not determine the standard of care for ethical advice, although a different standard would apply for advice in other areas. For example, a lawyer-ethicist would be held to a higher standard in advising about legal issues, but only to the standard of the average health care ethics consultant in advising about ethical issues.

Expectations of the Patient

Doctors who present themselves as specialists are held to a higher standard of care even if they are not specialists. For example, a chiropractor who represents that he/she possesses special skill and knowledge with regard to human ailments generally is held to the standard of a general practitioner.[26] It seems that, when determining the standard of care, the patient's reasonable perceptions and expectations of the process may govern the determination:

> But does the standard of skill to be applied depend on how the individual represents his or her ability or position or on what that position in fact is? This is an open question. In other words, a patient seeking treatment in a hospital emergency department who is ministered to by someone in a white coat with a stethoscope, appearing to be a physician, has certain expectations. If the individual turns out to have been an intern, what standard of care would be applied in a subsequent malpractice action? That of the "reasonable intern" in similar circumstances, or that of the "reasonable practitioner"? It could be argued that the reasonable expectations of the patient should govern in these circumstances.
>
> If a reasonable patient should have expected to be treated by a physician, then that standard must be applied. If, on the other hand, the intern is identified as such before attending to the patient and the patient authorizes the treatment, it could be said that a somewhat lower standard—that of the reasonable intern—ought to be applied.[27]

It can, therefore, be argued that a health professional without special training in ethics may be held to the standard of a health care ethics consultant, if the patient reasonably expected the health professional to have that special training. To the extent that expectations were raised in the patient, whether based on the hospital's literature, a health professional's remarks, or on the presentation of the health professional, these expectations might very well be reasonable and might therefore determine the standard of care. By the same token, if a health care ethics consultant represents that he/she has greater training, experience or qualifications than he/she in fact has, a court may hold him/her to the standard of care expected of the health care ethics consultant with that degree of experience.

Defenses in Negligence

In medical negligence law, there are two defenses that a health practitioner can raise to attempt to disprove a breach of the standard of care: error of judgment and approved practice. It is also open to other professionals to escape liability by showing that they were acting on their client's instructions.

ERROR OF JUDGMENT. A defense is open to health professionals to prove that, although they committed an error when considering the patient's case, they are not negligent because they possessed and exercised the skill, knowledge, and judgment of the average health professional of their special group. Canadian courts have accepted this defense:

> An error of judgment has long been distinguished from an act of unskillfulness or carelessness due to lack of knowledge. Although universally accepted procedures must be observed, they furnish little or no assistance in resolving such a predicament as faced the surgeon here. In such a situation, a decision must be made without delay based on limited known and unknown factors; and the honest and intelligent exercise of judgment has long been recognized as satisfying the professional obligation.[28]

In a more recent case, the Ontario High Court reiterated this principle:

> Needless to say, a mere error in judgment by a professional person is not by itself negligence. The courts recognize that professionals may make mistakes during the course of their practice, which do not bespeak negligence. Sometimes medical operations do not succeed. Sometimes lawyers lose cases. The mere fact of a poor result does not mean that there has been negligence. In order to succeed in an action against a professional person, a plaintiff must prove, on the balance of probabilities, not only that there has been a bad result, but that this has been brought about by negligent conduct.[29]

This defense would protect a health care ethics consultant who had met the standard of care, but an unfavorable result had occurred. If the patient or a family member believes after making the decision regarding treatment that it was a morally wrong decision and blames the health care ethics consultant for not having facilitated a better decision or for recommending the wrong course of action, this might constitute an unfavorable result. In such cases, the defense of error of judgment will protect any health care ethics consultant who honestly and intelligently turned his/her mind to an ethical problem with the competence of a reasonable health care ethics consultant in the same position.

This defense is particularly applicable to health care ethicists because the line between that which is morally wrong and right is often not perfectly clear, and because values can change over time. Since there is not an obviously right or wrong answer in many ethical problems, the defense of error of judgment may be invoked. What is involved is not a lack of skill so much as an exercise of the health care ethicist's judgment at the relevant time, and the court may be reluctant to say that choosing a certain ethical option was negligent. Certainly, if values have changed over time such that what was considered a morally right decision at the time is now considered to be morally wrong, the health care ethicist cannot be liable for the shifting values. The health care ethicist's advice must be evaluated according to the standards at the time it was given, not by a later, different standard.

APPROVED PRACTICE. It can be a defense to prove that the practice or procedure the medical professional followed was generally approved and employed by his/her colleagues at the time in issue. In *Goodwin v. Brady,* the British Columbia Supreme Court ruled that where there is a clear divergence of opinion on treatment of a medical problem, it is not negligence to adhere to one body of responsible medical opinion to the exclusion of another.[30]

The court in *Goodwin v. Brady* relied on the British case of *Bolam v. Friern Hospital Management Committee,* in which the court addressed the jury stating the general principles of the defense of approved practice:

> A doctor is not guilty of negligence if he has acted in accordance with a practice accepted as proper by a responsible body of medical men skilled in that particular art . . . Putting it the other way round, a doctor is not negligent if he is acting in accordance with such a practice, merely because there is a body of opinion that takes a contrary view.[31]

This defense is not absolute as the approved practice may be held to be inadequate. In *Gent and Gent v. Wilson,* the Ontario Court of Appeal stated:

> Each case must, of course, depend on its own particular facts. If the physician has rendered treatment in a manner which is in conformity with the standard and recognized practice followed by the members of his profession, unless that practice is demonstrably unsafe or dangerous, that fact affords cogent evidence that he has exercised that reasonable degree of care and skill that may be required of him.[32]

Although a departure from approved practice is a strong indication of failure to meet a reasonable standard of care, the effect of the defendant's conformity with approved practice is not certain. If the customary procedures are shown to be inadequate, the courts may find a professional to be negligent. A strong statement of this principle is seen in *Crits v. Sylvester:*

> Even if it had been established that what was done by the anesthetist was in accordance with "standard practice" such evidence is not necessarily to be taken as conclusive on an issue of negligence, particularly where the so called standard practice related to something which was not essential conduct requiring special medical skill and training either for its performance or a proper understanding of it . . . If it was standard practice, it was not a safe practice and should not have been followed.[33]

As can be seen from this excerpt, a court may exercise its powers to declare that a customary practice is inadequate more readily in regard to nontechnical matters. In the case of *Crits v. Sylvester,* the defendant anesthetist was found liable when an explosion occurred apparently as the result of static electricity and the anesthetist's use of a highly explosive mixture of ether and oxygen. The court found that the anesthetist had not given an explanation consistent with the absence of negligence. His conduct did not involve technical skill and experience, but was an omission to take proper precautions. Any sensible layperson would be competent to determine, without the assistance of expert evidence, whether such failure was negligence.

The court in *Crits* relied on a Manitoba judgment in *Anderson v. Chasney,* in which Mr. Justice Coyne stated:

> The opinions of the experts are not conclusive. But when an operation itself is a complicated and critical one, and acquaintance with anatomy, physiology

> or other subjects of expert medical knowledge, skill and experience are essential, jury or court may not be justified in disregarding such opinions and reaching conclusions based on views contrary to those of the experts. That is not the case here. Effective antecedent precautions were not taken and ordinary experience of jurymen or court is sufficient to enable them to pass upon the question whether such conduct constituted negligence.[34]

This could be significant in the context of health care ethics. For example, if a court were to see certain practical precautions as necessary before the health care ethics consultant gives advice, then, if these precautions are not taken, a court may well find a health care ethics consultant breached the standard of care, regardless of an approved practice that may not mandate these precautions. If these actions are seen to be common sense precautions, a court may decide that the health care consultant was negligent, without the help of expert testimony. Examples of precautions that may be deemed necessary include consulting the patient or examining the patient's chart. In any event, it might be extremely difficult for a defendant ethicist to establish an approved practice in a profession that is still in its infancy and that consists of members with various backgrounds who use different methodology.

Still, if a defendant ethics consultant can show that a significant proportion of his/her fellow professionals use the same methods and that these methods are not inadequate, he/she will be less likely to be found liable. To the extent that the ethics consultant can show that he/she has followed a recognized body of literature or of procedures, liability for negligence will be increasingly unlikely. For example, a health care ethicist who relies on materials outlining approved practice for the profession would probably be deemed to be acting reasonably. It must be remembered, however, that a court can always hold that the generally accepted practice is not in fact acceptable and that the health care ethicist should be liable.

CLIENT'S INSTRUCTION. In cases of legal malpractice, it is open to defendant lawyers to show that they were simply acting on their clients' instructions. Although it is the solicitor's duty to advise his/her clients of the legal hazards of pursuing an imprudent course of action, it is the clients' privilege to mismanage their affairs. Solicitors still must carry out their clients' instructions even if this entails not heeding their own advice. By analogy, as long as health care ethicists are not negligent in giving their consultation, if the physician or patient chooses to pursue an imprudent course of action, the results cannot be attributed to the health care ethicist.

Contributory Negligence

Any defendant found liable can reduce the damages recoverable by showing that the plaintiff was also negligent. If a plaintiff can be shown to have breached the standard of care expected of a reasonable person in his/her position, and if the breach is the factual and legal cause of his/her injury, then he/she is contributorily negligent. The standard of care expected of the reasonable patient is tied to the degree of knowledge of the layperson. A court may find, for example, that a patient who agrees to active euthanasia is contributorily negligent in not obeying society's moral and legal norms that prohibit such action. In addition, if a patient fails to disclose a material fact—a strong religious belief, for example—he/she might be found to be contributorily negligent.[35]

This means of reducing damages is rarely used in medical negligence cases. Courts may have traditionally set a low standard of care for patients since patients are usually in much weaker positions than physicians; they are often ill and submissive, and sometimes are incapable of looking after their own interests. As patients strive for a more equal and active role in their medical care, it is possible that more patients will be found contributorily negligent. Although it is not a strong defense at the moment, contributory negligence may be of increasing utility in reducing damage awards in the future.

The defense may be stronger when used against a physician instead of a patient. The standard of care attributable to a reasonable physician would probably be greater than the knowledge of a layperson, as physicians now receive at least some ethical training in medical school.[36] Courts may recognize that physicians do not occupy the same weak positions as patients with regard to health care ethics consultants; physicians are in a position to protect their own interests.

Although health professionals are generally entitled to rely on the opinion of specialists with regard to treatment within the specialist's field of expertise,[31] they are not entitled to delegate their own responsibilities. To the extent that a health care ethics consultant is considered to be a specialist in the field of ethics, with skills surpassing those of the ordinary health professional, the health professional should be entitled to rely on the advice received. It is doubtful, however, that a court would allow the health professional with overall responsibility for the care of the patient to escape liability entirely for an ethical decision. The health professional is likely to be found at least contributorily negligent.

The standard of care required of an ethics consultant will not often be affected by the knowledge or skill of the person advised. It will be affected only insofar as the reasonable expectations of the patient as to the health practitioner's degree of training and experience in ethics may raise the standard of care, as discussed under "Expectations of the Patient." Liability will therefore not often be affected by the characteristics of the plaintiff except with regard to the plaintiff's own action, which might be found contributorily negligent.

Plaintiff's Injury

A plaintiff will not be awarded damages in a negligence action unless he/she has suffered a material injury. In assessing damages, courts must look at the state of the patient prior to the negligent act and compare it with the state of the patient after the negligent act. The general principle for awarding damages caused by negligence is to restore the plaintiff to the position in which he/she would have been had the negligence not occurred. The goal is compensatory.

Clearly, in many situations in which a health care ethics consultant provides services, the state of the plaintiff prior to the consultation is not enviable. A court will not award damages to put the plaintiff in a position that would have been unattainable prior to the intervention of the health care ethicist.

Furthermore, there are certain categories of injuries in which the courts have been reluctant to award damages. Consider, for example, a "wrongful life" case in which a child attempts to sue the doctor or hospital in which the negligently performed abortion or sterilization procedure occurred that resulted in his/her birth. Courts have refused to award damages in "wrongful life cases" because they are not willing to assess the pre-negligence situation of the plaintiff for comparison purposes with the post-negligence situation. In suits brought by the parents in the same situations, courts have traditionally awarded only nominal damages.[38]

As discussed previously, damages recoverable under the Family Law Act are limited to expenses incurred as a result of the negligence and damages for loss of care, guidance, and companionship. A specific type of damages that may raise some concern is that of mental distress or mental suffering. Recently, the Ontario Court of Appeal in *Malette v. Shulman* upheld an award of $20,000 for mental distress against an emergency room physician.[39] The case involved the treatment of an unconscious victim of a car accident. The plaintiff patient carried a card identifying her to be a Jehovah's Witness and specifically refusing consent to a blood transfusion. The defendant physician disregarded the notice on the card and administered a blood transfusion because he considered the patient to be in a life threatening situation. When the patient recovered, she sued the hospital and attending doctors and nurses for battery and negligence. The court was quick to point out that the defendant physicians, hospital, and nurses were not negligent in the performance of their duties. Because the defendant physician who performed the transfusion had disregarded the informed refusal of the patient, however, he was liable in battery. Although the mental distress in this case resulted from a battery, the same principles to determine quantum of damages for mental distress would presumably apply if the court were to find a physician or ethicist negligent.

Causation

In order to establish negligence on the part of the health professional, the plaintiff must establish that the health professional's conduct was a cause in fact of the plaintiff's injury and that it was a proximate cause of that injury.

Cause in Fact

The legal test for cause in fact often used in medical negligence cases is the *sine qua non* or "but for" test. This test asks whether the injury would not have occurred but for the defendant's conduct. The Supreme Court of Canada has recently examined the test for cause in fact and has expanded cases in which cause in fact will be found. Mr. Justice Sopinka stressed that principles of proof should not be applied too rigidly and that causation is a practical question of fact that can best be answered by ordinary common sense, rather than abstract metaphysical theory.[40]

Despite the flexible standard of proof, cause in fact might be extremely hard for any plaintiff to prove against an ethics consultant. The physician or patient would have to prove that, but for the clarifications or recommendations of the ethics consultant, a different choice of treatment would have been made. Because the ultimate decision lies with the health professional, or the patient or the patient's family, a court might be reluctant to find that the health care ethicist's clarification, recommendation, or resolution of the issue is the cause in fact of any injury.

Courts have been willing to find that the medical professional who does not inform the patient of all the risks involved in the procedure is nevertheless not negligent if a reasonable patient would have consented to the treatment knowing all the material risks.[41] In such cases, cause in fact is not established. The onus of proving causation, like every other aspect of a negligence action, is on the plaintiff. If a health care ethics consultant does not adequately inform the patient or health professional, he/she could argue that a reasonable patient or health professional would still have followed the same course of action. Cause in fact would then not be established.

This may be the point in a negligence action where the position of the health care ethics consultant's advice on the spectrum between the problem clarification and problem resolution models might be important. If the ethics consultant gave an opinion or recommendation, or if he/she introduced an entirely new option that was eventually followed, this would be stronger evidence that the health care ethics consultant affected the treatment than if he/she merely clarified ethical issues. Cause in fact might be easier for a plaintiff to prove as the ethicist in question moves toward the problem resolution model.

Proximate Cause

Proximate cause or "remoteness" is a liability limiting device invented by the courts. It is a basic tenet of general negligence law that a defendant can only be liable for the consequences of his/her conduct that are foreseeable by a reasonable person in the defendant's position. The Supreme Court of Canada regards the reasonable foreseeability test as appropriate to determine the proximate cause in medical negligence cases. In *Cardin v. City of Montréal,*[42] the court stated that doctors should not be held responsible for unforeseeable accidents. In *University Hospital Board v. Lepine,* the Supreme Court of Canada wrote:

> Whether or not an act or omission is negligent must be judged not by its consequences alone but also by considering whether a reasonable person should have anticipated that what happened might be a natural result of that act or omission.[43]

A court would apply similar principles to an ethics consultant. If a physician or patient reacts to an ethicist's consultation in a manner unforeseeable to a reasonable ethicist, the ethicist would not be liable for the consequences of this reaction.

There is also a practical consideration that may well affect the decision of a court as to the proximate cause of a plaintiff's injury. The tendency of the courts to favor recovery by injured plaintiffs, especially where the defendants are insured, is not as apparent in cases of professional negligence.

> The temptation of diluting "fault" in favour of plaintiffs, evident in traffic and other cases involving insured defendants, has here been resisted, if only because of the adverse effect on professional reputation.[44]

Conclusion

The chances of a court finding a health care ethics consultant liable for damages are fairly small. The tort of battery, applicable in the medical context in certain specific contexts, is not likely to apply to a health care ethics consultant who only advises, clarifies the issues, or offers an opinion. Even in negligence, it is difficult to imagine many cases where a health care ethicist could cause actionable injury to a patient or physician.

Where a patient is harmed through medical treatment, it will be difficult for the plaintiff to prove that the health care ethicist's advice or clarification was the cause in fact of the injury. The patient, on whom the ultimate decision rests, will have to show that, but for the advice of the health care ethics consultant, he/she would not have undergone the treatment.

The defense of error of judgment should be a very strong ally for any health care ethicist. Health care ethics consultants will not be liable for their honest and intelligent exercise of judgment, as long as they employed the necessary skill, knowledge, and competence for their task. Neither will health care ethics consultants be liable for a decision that would have been seen as morally right at the relevant time, but has since become viewed as morally wrong because of a shift in moral values.

Furthermore, health care ethics consultants can attempt to absolve themselves of liability by showing that they were following the approved practice of their discipline. Of course, if a health care ethics consultant failed to follow certain procedures that a court deems necessary, such as consulting the patient's records, the health care ethics consultant may be held liable regardless of whether he/she was following generally accepted standards of practice.

The standard of care for an average health care ethics consultant will apply to all individuals who practice in this profession. Untrained consultants, who hold themselves out to be trained health care ethicists, will be held to the standard of the average health care ethics consultant. If they breach that standard, they will be held liable.

Of course, the standard of care to which health care ethicists will be held is unpredictable, but if the health care ethics consultant follows the basic methodology of his/her discipline and uses the knowledge and skills of the average health care ethics consultant in the profession, it is unlikely that a court would find that the health care ethics consultant was negligent in the course of his/her professional duties.

Notes and References

[1]This chapter was prepared with the assistance of Laura Fric, summer student at Osler, Hoskin & Harcourt in 1992. The discussion in this chapter is necessarily of a general nature and cannot be regarded as legal advice. The authors will be pleased to provide additional details on request and to discuss the possible effect of these matters in specific situations.

[2]179 Cal. App. 3d 1127; 225 Cal. Rptr. 297.

[3]Colleen Clements, "Enlisting Aid of Bioethicist Can Be Double-Edged Sword," *Medical Post* 21 May 1991: 30.

[4]John A. Robertson, "Clinical Medical Ethics and the Law: The Rights and Duties of Ethics Consultants," *Ethics Consultation in Health Care,* eds. John C. Fletcher, Norman Quist, and Albert R. Jonsen (Ann Arbor, MI: Health Administration, 1989) 165,166.

[5]Donnie J. Self and Joy D. Skeel, "Legal Liability and Clinical Ethics Consultants: Practical and Philosophical Considerations," *Medical Ethics: A Guide for Health; Professionals,* eds. John Monagle and David C. Thomasma (Rockville, MD: Aspen, 1988) 408–416; Donnie J. Self and Joy D. Skeel, "Professional Liability (Malpractice) Coverage of Humanist Scholars Functioning as Clinical Medical Ethicists," *Journal of Medical Humanities and Bioethics* 9 (1988): 101–110.

[6]*See Hopp v.Lepp* [1980] 2 S.C.R. 192; 13 C.C.L.T. 66 and *Reibl v. Hughes* [1980] 2 S.C.R. 880; 14 C.C.L.T. 1.

[7]John G. Fleming, *The Law of Torts,* 6th ed. (Sydney: Law Book, 1983) 23.

[8]*Reibl v. Hughes.*

[9]*Reibl v. Hughes,* at C.C.L.T. p. 14.

[10]*Zimmer v. Ringrose* (1978), 16 C.C.L.T. 51 at p. 56.

[11]*Piggly Wiggly v. Rickles,* 212 Ala. 585, 103 So. 860.
[12]*Green v. Goddard* (1702), 2 Salk. 641, 91 E.R. 540.
[13]*Dodwell v. Burford* (1669), 1 Mod. 24, 86 E.R. 703.
[14]*Innes v. Wylie* (1844), 1 C. & K. 257.
[15]This chapter will continue to discuss the liability of a health care ethics consultant and all remarks will apply equally to a health care ethics committee unless otherwise stated.
[16][1932] A.C. 562 at 580.
[17]Arthur J. Meagher, Peter J. Marr, and Ronald A. Meagher, *Doctors and Hositals: Legal Duties* (Toronto: Butterworths, 1991) 179.
[18]Robertson (166–168), comes to this conclusion and goes on to examine the potential legal duties of ethicists.
[19]R.S.O. 1990, c. F.3.
[20](1956), O.R. 132 at p. 143.
[21](1972), 28 D.L.R. (3d) 521, aff'd in part (1973), 41 D.L.R. (3d) 321 (Ont. C.A.), aff'd (1976), 66 D.L.R. (3d) 530 (S.C.C.).
[22]Michael D. Coughlin and John L. Watts, "Ethics Consultation in Canada: An Empirical Study," Money, Power and People: Social Dimensions of Bioethics, 4th Annual Meeting of the Canadian Bioethics Society, Toronto, October 31, 1992.
[23]Robertson 166.
[24]*Rietze v. Bruser* (No.2), [1979] 1 W.W.R. 31 at 45 (Man. Q.B.).
[25]Ellen Picard, *Legal Liability of Doctors and Hospitals in Canada,* 2nd ed. (Toronto: Butterworths, 1984) 156,157.
[26]*Gibbons v. Harris,* [1924] 1 D.L.R. 923 (Alta. C.A.).
[27]Gilbert Sharpe, *The Law and Medicine in Canada,* 2nd ed. (Toronto: Butterworths, 1987) 19,20.
[28]*Wilson v. Swanson* (1956), 5 D.L.R. (2d) 113 at 120, per Rand, J.
[29]*White v. Turner* (1981), 120 D.L.R.(3d) 269 at 278.
[30](1991), 7 C.C.L.T. (2d) 319 (B.C. S.C.).
[31][1957], 2 All E.R. 118 (Q.B.), at p. 122.
[32][1956], O.R. 257 at 265.
[33][1956] O.R. 132 at 150 (Ont. C.A.), aff'd [1956] S.C.R. 991 at 997.
[34][1949] 2 W.W.R. 337, aff'd [1950] 4 D.L.R. 223 (S.C.C.), at p. 359.
[35]Picard 248.
[36]The extent of ethical training varies considerably among medical schools. *See* Françoise Baylis and Jocelyn Downie, *Undergraduate Medical Ethics Education: A Survey of Canadian Medical Schools* (London: Westminster Institute for Ethics and Human Values, 1990) 30.
[37]*Malette v. Shulman* (1987), 47 D.L.R. (4th) 18, aff'd (1990) 67 D.L.R. (4th) 321.

[38]*Doiron v. Orr* (1978), 20 O.R. (2d) 71; *Cataford v. Moreau* (1978), 7 C.C.L.T. 241.
[39](1990), 67 D.L.R. (4th) 321.
[40]*Snell v. Farrell* (1990), 72 D.L.R. (4th) 289 at p. 300.
[41]*Videto v. Kennedy* (1981), 33 O.R. (2d) 497, 17 C.C.L.T. 307 (C.A.); *Zimmer v. Ringrose* (1981), 16 C.C.L.T. 51, leave to appeal to S.C.C. denied.
[42](1961), 29 D.L.R. (2d) 492 at 494.
[43][1966] S.C.R. 561 at 579.
[44]Fleming 105.

What Does a Health Care Ethics Consultant Look Like?

Results of a Canadian Survey

Michael D. Coughlin and John L. Watts

When the Network on Health Care Ethics Consultation began developing "A Profile of the Health Care Ethics Consultant," it quickly became apparent that information was needed about who was actually doing ethics consultation, what kind of training and skills they possessed, what ethics activities they engaged in, and what their views were on certification. Such knowledge would serve as a "reality check" on the deliberations of the Network and would help focus the discussions.

A survey of the relevant literature revealed very little empirical data on the topic.[1] Although there is work in progress to create a comprehensive data base of health care ethics consultants in North America,[2] and there is a brief report of a survey of ethics consultants covering data between 1980 and 1984,[3] we found only one descriptive study on ethics consultation in the United States. This study by Donnie J. Self and Joy D. Skeel,[4] however, is limited in several respects. First, the report focuses narrowly on legal liability, documentation, and accountability. Second, the report looks only at "humanist" ethics consultants; that is, consultants with backgrounds in philosophy, theology, or other university disciplines in the humanities. Physicians, nurses, and other health care workers who provide ethics consultation are excluded from consideration, and this constitutes a serious limitation since the report cannot then adequately reflect the reality of current practice. Third, the study is dated; it was completed more than five years ago, and the field of ethics consultation has changed rapidly in the intervening period.

Given the above, it was important to conduct a current and more comprehensive survey of those doing ethics consultation in health care. The survey reported here provides a comprehensive overview of the current state of ethics consultation within Canada, and expands on a preliminary report of this survey published previously.[5] Although the population studied is limited to Canada, it is reasonable to believe that the findings in this study may be applicable to ethics consultants throughout North America.

Methodology

Development of the Questionnaire

The first task was to define the term "health care ethics consultation." For the purposes of the questionnaire, health care ethics consultation was defined broadly to include consultation on ethical issues in clinical cases or in clinical research, ethics consultation to ethics committees, research ethics boards (institutional review boards), or policy formulation committees in health care institutions. Since there is no standard or universally accepted definition of the term, and since an important objective of the study was to determine the degree of variation among those who consider themselves to be providing ethics consultation, a broad definition of the key term seemed appropriate.

The questionnaire was developed on the basis of questions that arose in discussing "A Profile of the Health Care Ethics Consultant." Suggestions and criticisms of pilot drafts by members of the Network shaped the final instrument. The questionnaire contained 40 questions that examined demographics, educational background, time spent on ethics, approach to the role of consultation, research-related issues and, finally, attitudes toward certification (*see* Appendix). Most questions had predetermined, multiple-choice alternatives for answers, but some questions allowed for open-ended responses to capture as much diversity as possible. Intelligibility and face validity of the questionnaire were established by having members of the Network complete the final draft version.

Compilation of Mailing List

The methodology for distribution of the questionnaire has been reported.[6] Since the goal of the survey was to be as inclusive as possible, a number of different sources were used to compile a mailing list of 350 individuals and institutions:

1. Persons who responded to a request for self-identification made at the 1991 Annual Meeting of the Canadian Bioethics Society.
2. Members of the Canadian Bioethics Society identified by member of the Network as possible providers of ethics consultation.
3. Members of the ethics committee Network of South Central Ontario.[7]
4. Contact persons on ethics committees in Québec obtained from a 1991 study conducted by the Group de Recherche en Éthique Médicale at Laval University.[8]

5. Hospitals in Canada with over 300 beds.
6. All children's hospitals in Canada.

Although even a list from such diverse sources will not be all-inclusive, we believe that the overwhelming majority of those providing ethics consultation in Canada were included in the list. And because the questionnaire was sent both to individuals and to institutions, we expected that "ethics consultants" not on the lists of individuals would be contacted through institutions. This approach necessarily involved a good deal of overlap and we expect that this duplication was responsible for at least some of the nonresponses.

Distribution and Collection of Data

Number-coded questionnaires in both French and English were sent to all individuals on the mailing list with a cover letter explaining the purpose of the study. To increase the response rate, a second "reminder" letter was sent out just before the deadline date, with an opportunity for those who had lost or not received the questionnaire to respond. Recipients were asked to return the questionnaire even if they did not provide health care ethics consultation.

To maintain confidentiality, names of respondents did not appear on the questionnaire. Returned questionnaires were sent to the bioethics department at The Hospital for Sick Children (Toronto). The receiving person noted the numeric code on a list and then discarded the envelope. Questionnaires were bundled and sent to the ethics department at St. Joseph's Hospital (Hamilton) for compilation and analysis. This was handled through the Computation Services Unit at McMaster University using the SPSS program. No statistical analysis of the data is included because such analysis is not applicable to studies such as this one which looks at essentially the whole population of health care ethics consultants in Canada and not just a sampling.

Table 1
Response to Survey

Questionnaires mailed	350
Responses received	253
Response rate	72%
Questionnaires completed (respondents)	161
Not providing ethics consultation (as individuals)	83
Returned address unknown	8
Duplicate response	1

Table 2
Regional Representation

Region	Population, % of total	Sent		Completed	
Atlantic	10%	21	(6%)	9	(6%)
Québec	25%	67	(19%)	32	(20%)
Ontario	37%	188	(54%)	97	(60%)
Western	28%	74	(21%)	23	(14%)
	100%	350	(100%)	161	(100%)

Results

Response Rate

The total response rate from the 350 questionnaires sent out was 72% (253 questionnaires), and 64% of that group (161 out of 253) indicated that they were, to some extent, ethics consultants (Table 1). These 161 questionnaires serve as the basis for this analysis. These individuals are called "respondents" in the remainder of this report.

Demographics

Although the regional distribution of the questionnaires sent out did not match regional population differences, the distribution of questionnaires returned was similar to that of those sent out and did represent the whole country, although not proportionally (Table 2). Ontario was

Table 3
Age, Sex, and Language of Respondents

Age		
20–29	1	(.5%)
30–39	32	(20%)
40–49	50	(31%)
50–59	52	(32%)
60+	25	(16%)
No response	1	(.5%)
	161	(100%)
Sex		
Male	112	(70%)
Female	45	(28%)
No response	4	(2%)
	161	(100%)
Language		
English	138	(86%)
French	23	(14%)
	161	(100%)

somewhat overrepresented in the survey. Nevertheless, although the proportion of respondents relative to the population was greater in Ontario than in Québec, the relative proportions of those spending 30% or more of their time was equivalent in both provinces: 13 persons in Québec and 21 persons in Ontario. Thus, in this group, Québec had two-thirds the number of consultants as Ontario, corresponding to a total population two-thirds that of Ontario.

The great majority of individuals doing health care ethics consultation were over 40 years of age, with only one respondent under 30 years of age (Table 3). In fact, about half the respondents were 50 years of age or older. This may result from the fact that most individuals come to ethics consultation after prior training and experience in some other discipline or profession. A minority, only 28%, of those doing ethics consultation are female.[9] The 23 francophones,

Table 4
Factors Determining a "Bona Fide" Ethics Consultant[a]

Education/training	46
Specific postgraduate degree in philosophy/ethics/theology	18
Ethics is primary commitment/employment	15
Experience	10
Receives payment/salary for ethics consultation	8
Peer recognition	4

[a]97 respondents—some of whom identified more than one determining factor.

while only 14% of the total respondents, accounted for 75% of the respondents from Québec.

Determining Who Is a Health Care Ethics Consultant

The questionnaire invited respondents to define/describe a "bona fide" health care ethics consultant (Table 4). Of the 161 individuals who completed the questionnaire, 97 responded to this question and their answer can be grouped around the following themes:

1. Education and training
2. Employment and time commitment.
3. Peer recognition and research.
4. Role and function.

Education and Training

A majority (46 out of 97) gave answers that emphasized a requirement for previous education or training, with 18 of the 46 specifically stating that this should be at the master's or PhD level in philosophy or theology (e.g., "individual with PhD in ethics," "doctoral degree in philosophical or theological ethics.") Several respondents wished to include both training and work experience in the description: "a person with formal training in bioethics at the master's or doctoral level who has clinical experience in a health care setting." Others placed greater emphasis on recognized current competence, such as "a person with recognized ability to provide a reasonable approach to the moral component of medical practice and research," "one who has the training and skill to recognize and analyze ethical issues in clinical areas and can articulate this in a way that promotes good decision making" or more simply, "expertise through training and experience." Suggestions range from the highly specific "advanced degree in theology and philosophy; clinical exposure 3+ years and practicing committee work for 3 years," to the highly general "professionally qualified and competent." Finally, there were the iconoclasts, ranging from the individual who answered "someone like me" to the somewhat insulted "this is a trade union question which is self-serving on the part of those circulating the questionnaire."

Several years ago, John La Puma and David L. Schiedermayer argued that health care ethics consultants should ideally be "clinicians who are expert in their own medical discipline . . . "[10] In their article, the authors note that there is a controversy over whether health care ethics consultants must be physicians. Respondents in our survey often included reference to the need for clinical experience: "academic preparation; hands-on clinical involvement; multidisciplinary openness." However, only one of the respondents made a statement that might be interpreted as requiring the ethics consultant to have prior credentials in medicine: "a person well trained in clinical ethics—requiring philosophical and legal knowledge and a good practical grounding in clinical medicine." What the comments did highlight in regard to preparation was the need for training in ethics (primarily academic), made effective and practical through clinical experience, or as one respondent put it, "a person with sufficient formal or informal training in ethics consultation who is skilled at performing the required tasks." Later in this chapter, we examine the actual education and training of those who currently identify themselves as health care ethics consultants.

Employment and Time Commitment

Fifteen of the 97 defined a health care ethics consultant as an individual who was specifically employed by an institution to do such work. A further eight respondents mentioned that the consultant was paid for the clinical ethics work. Associated with the employment aspect was usually a comment on the time spent: "full time employed by hospital and university," "full time or major part-time and funded specifically," "salaried position including significant expectation of time to be spent in clinical consultation," and "someone who devotes at least 50% of his/her time to ethics matters." Thus, payment was usually linked to an expectation of a major time commitment. In fact, as we show below, time commitment was a crucial factor in analyzing the results of our survey. For example, specific employment will often include a title for the position. However, it appears that use of a title will not necessarily identify a "bona fide" health care ethics consultant if other requirements are also taken into account. As detailed in our previous report, among the 43 respondents using some sort of title (e.g., ethics consultant, clinical ethicist, or the like), 13 spent less than 10% of their time in clinical ethics work, and 10 were not paid at all for this work.

Peer Recognition and Research

One of the other criteria mentioned by a number of respondents was peer recognition. An ethics consultant is a "person with recognized ability," who is "recognized to have expertise." Although not always linked to peer recognition, research and publication are also a type of peer recognition, and they were noted by a number of respondents. We examine below the important role academic pursuits play in the activities of ethics consultants, especially for those who spend a high proportion of their time in clinical ethics work.

Role and Function

In comments on the "bona fide" health care ethics consultant, respondents mentioned roles, such as "to provide staff with guidance and education on ethical issues . . . " The ethics consultant is "one who seriously walks through ethical decision making and analysis processes with people who ask for help and/or who assists with development and evaluation of policies, practices and research with attention to the ethical dimensions of these." Elsewhere in the questionnaire, when asked what they considered their own most important function as ethics consultant, the largest percentage (21%) said that "education" was their most important function. Next came "clarification of issues" (17%), followed by "widening perspectives and appreciation of alternatives" (10%), and "enhancing awareness of ethical issues" (8%).

Using their own criteria, most of the respondents would not meet the definition of "bona fide" health care ethics consultant. However, they recognize this and prefer to identify themselves as doing ethics consultation rather than being ethics consultants.

Education and Training

Respondents were asked to list all their professional and academic qualifications and the year in which they were obtained. After the Network had identified feeder disciplines and qualifications (*see* chapter on "Feeder Disciplines: The Education and Training of Health Care Ethics Consultants"), these responses were categorized according to primary, secondary, and tertiary (for a few) feeder qualifications. A primary feeder qualification was defined as the first degree acquired that would provide entry into that particular feeder profession, e.g., MD or equivalent for medicine, PhD for philosophy, but BScN or equivalent for nursing.

Few ethics consultants are trained specifically for this role; only 24 respondents had some kind of degree specialization in bioethics. Thirteen of these were at the doctoral level in philosophy or theology, nine at the master's level in philosophy, theology or law, and two others noted a bioethics specialization without degree. Most with degree specialization were English speaking, with only three of the 24 being francophone. Those with specialization in ethics tended to be younger than the total group, with 42% (10) being under 40 years of age, and another 29% (7) between 40 and 49 years of age. In contrast, only 20% of all the respondents are under 40 years of age, whereas the 40–49-year-old group accounts for 31% of total respondents. There is also a gender difference. Whereas women account for only 28% of total respondents, they constitute 42% (10) of the group with a specialized degree in bioethics.

Approximately half of the respondents came from a health care feeder discipline: 32% from medicine, 12% from nursing, and 9% from other health and related professions. The other half came predominantly from the normative disciplines: 22% from theology, 15% from philosophy, and 8% from law (Table 5).

The levels of training for the feeder disciplines represented a generally advanced level of education. In the normative disciplines, 62% had a doctoral degree (23 of 24 in philosophy, 17 of 36 in theology, and 5 of 13 in law). In the health care professions, 6 of the 19 nurses had a master's degree in nursing and 6 had a doctorate in a secondary area of specialization.[11] If one assumes that all physicians can be considered to have doctoral degrees, a total of 109 persons had doctoral degrees in their entry profession. There were a further additional 11 doctorates among the second degrees and a total of 116 respondents (72%) had a doctorate, 4 of those respondents having two doctoral degrees.

Table 5
Academic Qualifications of Those Doing Ethics Consultation

Discipline	Discipline represented in total group of respondents		Primary qualification of respondents[a]	
Medicine	53	(33%)	52	(32%)
Nursing	19	(12%)	19	(12%)
Philosophy	54	(34%)	24	(15%)
Theology	42	(26%)	36	(22%)
Law	17	(11%)	13	(8%)
Other[b]	75	(47%)	15	(9%)
No response			2	(1%)
			161	(99%)[c]

[a]Primary qualification is defined as the first qualification achieved that would constitute entry into a "feeder discipline" (e.g., LLB for law, MB for medicine).
[b]Includes health administration, education, science, psychology, and so on.
[c]Total percentage does not equal 100 because of rounding.

The median year in which physicians and nurses doing ethics received their degrees was 1968, corresponding to the belief that most of the people doing ethics consultation are already well established in their careers. However, those in the allied health professions were somewhat younger in their career with the median year of professional qualification being 1982. Respondents in the normative sciences tended to be somewhat younger than physicians and nurses with the median year of the feeder degree being in the early to mid-1970s.

Multidisciplinarity

In our previous analysis of this data, we reported that a high proportion of respondents had degrees or training in disciplines other than that of their primary qualification (Table 5). In fact, of those who spent 30% or more of their time in clinical ethics work (*see* discussion below of high and low proportion groups), 51% had at least one degree in philosophy and 33% had a degree in theology, with an overlap of individuals who had both. This suggested that

Table 6
Distribution of Multiple-Degree Competencies

Primary degree	Secondary degree						
	Med'n	Nursg.	Phil.	Theol.	Law	Other	Total
Medicine	0	0	3	0	2	0	5
Nursing	0	0	3	3	0	7	13
Philosophy	0	0	0	1	0	1	2
Theology	0	0	1	0	1	1	3
Law	0	0	2	1	0	0	3

those involved in ethics consultation may have training in several areas of competence. In fact, 26 of the respondents had a second degree that would represent qualification for a second feeder profession (Table 6). Four of the 26 had qualifications that would represent entry criteria for three different feeder disciplines (e.g., nursing, law, and philosophy). Of the 26, 12 spent more than 30% of their time in ethics, or put alternatively, 28% of those in the high proportion group had double degrees compared with 11% in the low proportion group. The largest single group having double degrees was nursing with three in theology, two in philosophy, and seven in the "other" category. However, this is understandable because, for the purpose of this survey, entry qualification for nursing was usually a first degree whereas for the other professions, it was a second or postbaccalaureate degree. Although there seems to be significant multidisciplinary training among those doing ethics consultation, we do not have the data to determine if this is different from the profile of health professionals who are not involved in ethics consultation.

Training in Health Care Ethics

Most of those who were doing ethics consultation had no degree specialization in ethics. Even among those spending a high proportion of their time in ethics work, only 40% had a degree specialization in bioethics, although many

Table 7
Nonspecialist Training in Health Care Ethics

For whole group[a]		For high-proportion group[b]	
Educational experience		Educational experience	
Member of ethics committee	51	Member of ethics committee	8
Self-directed reading	42	Seminars/workshops	8
Seminars/workshops	41	Teaching bioethics	6
Courses	21	Courses	4
Teaching bioethics	20	Chair of REB (IRB)	4
Research and publication	16	Self-directed reading	4
Chair of ethics committee	14	Consultation experience	4

[a]Represents responses from 98 (or 61%) of the 161 respondents. Of the remainder, 25 had a specialized degree in bioethics and would not have responded to this question. Respondents gave up to three different kinds of educational approaches.

[b]Represents responses from 18 (40%) of the 43 individuals in the high-proportion group. Note that 17 of the 43 had bioethics specializations and would not be included here, and another 17 had some formal training in ethics in addition to the bioethics training noted here.

had backgrounds in philosophy, theology, or law. We asked respondents how they had acquired sufficient expertise to provide ethics consultation. The results are shown in Table 7. For those with no specialized training in bioethics, the most commonly cited route of training was membership on an ethics committee. Even for those in the group pending a high proportion of their time in ethics work, membership on an ethics committee was considered to be a prime learning experience in the absence of specialized training. This points out the importance of the ethics committee as a training ground for those involved in ethics consultation in health care. The other major sources of training were seminars and workshops, self-directed reading, and active involvement in ethics through teaching or other experience. For many doing ethics consultation, therefore, ethics learning tends to be primarily self-directed and by "apprenticeship."

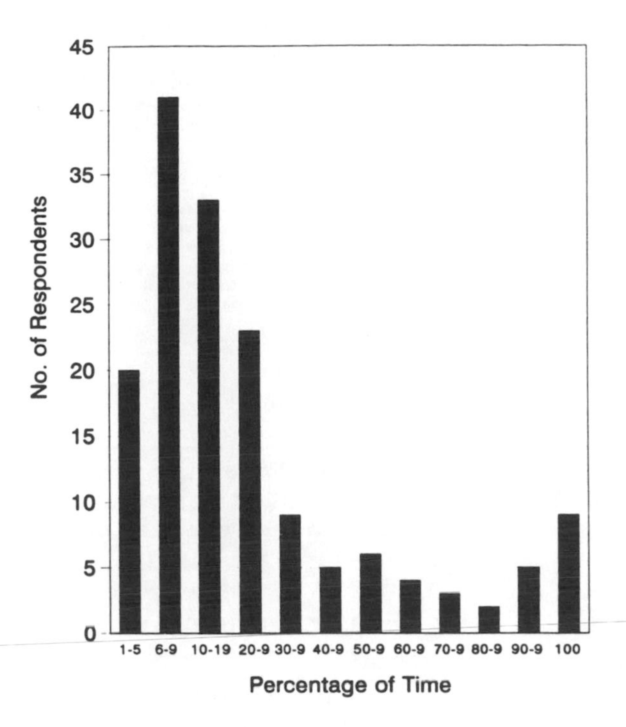

Fig. 1. Percentage of time spent in clinical ethics work.

Employment

Of the respondents, 12 were full-time ethics consultants, and 24 were part-time consultants. However, while the full-time consultants reported spending 40 to 100% of their time on clinical ethics work, part-time consultants ranged from 7 to 100% for time spent on clinical ethics work. In fact, among all the respondents, there was a great deal of variability in the amount of time spent on clinical ethics work (Fig. 1). In our previous report, using this data, we identified two groups of respondents, one that spent less than 30% of their work time doing clinical ethics work (the "low proportion group"), and the other that spent 30% or more of their time in clinical ethics work (the "high propor-

Table 8
Relationship of Primary Qualification to Time Spent in Ethics

Background	Low-proportion group[a]		High-proportion group[b]	
Medicine	47	(41%)	5	(12%)
Nursing	12	(10%)	6	(14%)
Philosophy	10	(9%)	14	(33%)
Theology	26	(23%)	10	(24%)
Law	7	(6%)	6	(14%)
Other	13	(11%)	1	(2%)
Total	115	(100%)	42	(99%)[c]
No response 4 (2%)				

[a]Spend less than 30% of time in clinical ethics work.
[b]Spend 30% or more of time in clinical ethics work.
[c]Total percentage does not equal 100 because of rounding.

tion group"). The high proportion group included all those who identified themselves as full-time ethics consultants and half of those who identified themselves as part-time ethics consultants. The makeup of the high proportion group was somewhat different from that of the low proportion group: 33% of the high proportion group were less than 40 years of age compared to 17% of the low proportion group. The ratio of women to men was also somewhat different with 35% of the high proportion group being female compared to 25% of the low proportion group. Those spending a high proportion of their time on clinical ethics work come primarily from the normative disciplines, whereas the health professionals tend to spend less time on ethics (Table 8).

Forty-four percent of individuals providing ethics consultation reported that this was part of their job description.

That leaves a significant proportion for whom providing ethics consultation is done on a voluntary basis. The high degree of voluntary consultation is confirmed by the fact that only 28% of those in the low proportion group (spending 30% of time on clinical ethics work) reported receiving any payment for providing ethics consultation.

Table 9
Sources of Consultation Requests

Clinical ethics committee	75%
Physicians	75%
Nurses	59%
Other health professionals	51%
Hospital administration	38%
Patients/families	29%

In contrast, 91% of those in the high proportion group received some payment.

Research

Academic pursuits play an important role in the activities of respondents, despite the fact that slightly fewer are employed primarily by universities or bioethics centers (36%) than by hospitals (42%). At least 60% regard themselves as being actively involved in research, with 29% receiving protected time for this activity, and 21% having held grants as principal investigators. Fifty-two percent of the respondents have published scholarly articles, and 15% have published or contributed to books.

The Consultation Process

Only 22% of the respondents consult as either full- or part-time consultants; the majority (79%) act as members of an ethics committee, whereas a few do consultations as part of another primary role, such as hospital chaplain. It is not surprising then that three quarters of these respondents identified ethics committees as being the instigators of consultations. However, the majority received requests from multiple sources (Table 9). Physicians were the most frequent individual sources (75% of respondents received requests from this source), followed by nurses (59%), and other health professionals (51%). Only 29% of consultants actually expected and received requests from patients and their families.

Table 10
Types of Health Care Ethics Consultation Provided

Clinical case consultation	78%
Policy formulation on ethical issue	72%
Consultation to ethical committees	62%
Clinical research consultation	39%
Consultation to REB (IRB)	37%
Consultation to other ethics consultants	22%

Just as the sources of consultation vary and are multiple, so too the nature of consultations varies, and most consultants receive requests in several different areas. Clinical case consultation continues to dominate (with 78% of respondents acting in this way), but is followed closely by consultation on policy issues (72%) (Table 10). This means that 22% of respondents do not do individual case consultations. They may regard themselves as responding to clinical ethics committees or to research ethics boards on policy issues.

Clinical Case Consultation

The frequency of clinical case consultation is relatively uniform throughout the various groups. Among all consultants, the median number of consultations per month is two (with a range of 0–20), and this frequency is independent of the percent of time spent on clinical ethics work. The feeder disciplines of theology and law have a slightly higher median (three case consults per month); medicine, nursing, and philosophy have a median of two per month; and the "other" feeder group was somewhat lower (one per month). Despite the frequency of specific case consultation, it appears that ethics consultation may not have acquired the legitimacy of its medical counterpart, since only 15% of consultants report that a written summary is always placed in the chart, whereas 38% say this never or rarely happens (Table 11). Even when a summary does

Table 11
Written Records of Clinical Case Consultation

Written summary	Always	Often	Sometimes	Rarely	Never
Prepared and filed	19%	12%	30%	9%	21%
Put in patient chart	15%	12%	19%	9%	29%

appear in the chart, it is as likely to be written by the attending physician as by the consultant. A written summary that does not accompany the chart is not as infrequent as a note in the chart. Whether this means that ethics consultations are not yet regarded seriously by hospital administration or by physicians is not known. In at least some centers, the lack of a formal record results from fear of liability either on the part of the institution or possibly on the part of the ethics consultant.[12]

Given the focus on clinical cases, and the fact that most respondents work in acute care hospitals, we find among the types of issues for which consultations are sought the predominance of informed consent, withdrawal of intensive care or life support, and surrogate decision making (Table 12). Also, issues related to the allocation of resources are increasingly common, although triage as a specific concern is rarely mentioned.

Among issues for which consultations are infrequently requested are those of child or elder abuse and the nature of appropriate psychiatric treatment. Perhaps these issues are not being brought to ethics consultants because clinicians consider them to be more related to legal rather than ethical concerns. The absence of the use of restraints as a common cause for concern probably reflects the background institutions of the respondents.

Termination of pregnancy is also rarely considered by ethics consultants. There are several possible reasons for this, including the possibility that clinicians may themselves hold extremely strong views on the issue in one

Table 12
Ranking of the Most Common Issues For Clinical Consultation

Ethical issues[a]	Rank of issues by times cited[b]		Rank of issues by times cited[b] as most common[c]	
Withdrawal of treatment	1.	71%	1.	43%
Informed consent	2.	64%	2.	33%
Refusal of treatment	3.	56%	5.	27%
Care for dying patients	4.	54%	4.	28%
Surrogate decision making	5.	54%	3.	39%
Confidentiality	6.	52%	6.	26%
Nutrition and hydration	7.	48%	9.	10%
Living wills/advance directives	8.	43%	10.	10%
Allocation of resources	9.	43%	8.	20%
Interprofessional conflicts	10.	40%	15.	7%
Request for futile treatment	11.	39%	12.	8%
Clinical research-related issues	12.	38%	7.	22%
Family conflicts	13.	35%	16.	7%
Euthanasia	14.	34%	13.	9%
HIV-related issues	15.	33%	17.	7%
Truth-telling	16.	33%	19.	6%
Competency evaluation	17.	30%	14.	7%
Reproductive issues	18.	28%	11.	6%
Use of restraints	19.	27%	21.	<5%
Organ donation	20.	22%	23.	<5%
Discharge of patients at risk	21.	20%	20.	<5%
Appropriate psychiatric treatment	22.	18%	22.	<5%
Termination of pregnancy	23.	16%	24.	<5%
Triage within/across institutions	24.	14%	25.	<5%
Child or elder abuse	25.	11%	26.	<5%
Other	26.	7%	18.	6%
Animal experimentation	27.	6%	27.	<5%

[a]Issues appearing in the questionnaire are presented here in their rank ordering of most to least commonly associated with clinical case consultations.

[b]Respondents were asked to check off all issues that were common as well as the three (3) that were most common. This ranking is by the percentage of respondents that cited the issue either way.

[c]For comparison, the rank ordering of the issues is given by the percentage of those citing the issue as one of the three most common in their practice. This ordering is very similar to the previous.

direction or the other and, thus, are not interested in seeking consultation. They may, on the other hand, feel that the degree of social and political disagreement on the topic limits the effectiveness of consultation.

Also noteworthy is the relative infrequency of discussion of issues related to organ donation and those related to infection with human immunodeficiency virus (HIV). It is possible that the low profile of the latter arises from the relatively lower incidence of AIDS and HIV in Canada, and the fact that its distribution results in a relatively high prevalence in a very small number of large cities and a very low prevalence elsewhere in the country.

Certification of Ethics Consultants

Health care ethics consultants in Canada appear to be evenly divided on the advisability of formal certification for ethics consultants, with 43% of respondents in favor and 40% against. This split was seen consistently and appeared to be independent of time spent in consulting practice. Even those individuals who spent more than 30% of their time practicing ethics consultation were evenly divided on this issue.

Among those who are not in favor of certification, several points emerged. To some people, the apparent absence of incompetent practitioners or charlatans was reassuring: "I am not aware of any abuses in this area. Someone would have to provide evidence to convince me that there is a need for certification. In the absence of evidence of a need, I think, a move toward certification is very ill advised." For others, the common practice of consultation in the presence of a committee structure was seen as sufficient defense: "self-directed learning, experience and leadership within the committee should offset the risk of incompetent individuals giving advice on ethical issues." The absence of commonly accepted educational and training prerequisites was seen not only as a practical problem for accreditation/certification, but was viewed as a positive strength, leading to flexibility and desirable heterogeneity: "medical ethics must resist efforts of homogenization," "value the diversity offered by each discipline which allows teams to draw on resources that can supplement their strengths and fill the gaps that they have as a group." The need to maintain such diversity was perhaps the strongest and most frequent comment made by those not in favor of certification.

Many of those who did not favor certification, however, did so mainly because "the field is not yet sufficiently developed" and identified a potential future need—"the need is growing but I am not sure we are there yet." Similarly, those not in favor recognized that guidelines for training and an expectation of education, at least at the master's level, should be encouraged.

There were some who were very positive about the need for certification:

> Many people trained in the professions (doctor, lawyer, clergy) imagine that a special interest in ethics automatically renders them equipped for the task. Some seem to think that good will and a few seminars in ethics are sufficient. I have extensive postgraduate training in ethics but I do not call myself a clinical ethicist and do not believe I should until I have a Ph.D. I believe that clinical ethics is a specific professional field which needs more accountability in the form of professional associations.

This view is typical of those who believed certification would also require doctoral training in philosophy or theology. However, a larger number approved certification, either on a voluntary basis only, or for those practicing "professional" or full-time directive consultation only. Typically, those approving certification did so with a sense of ambiguity: "I do not really like the idea of certification but there must be some way of allowing for evaluation of training, perhaps by an apprenticeship, fellowship or the like."

"Common sense and experience count for a lot but I believe formal preparation, philosophy and/or theology must be a part of the package." "There is a need for control to curb the activities of self-styled "ethicists" but I would not like to see rigid rules either. We already have too many professional boxes which promote defensiveness and prevent team work and collaboration." Many people on both sides of the question pointed out the need to avoid outlawing the committed, volunteer, part-time ethics consultant (although, not surprisingly, there is a small number for whom that committed volunteer is just as likely to be regarded as a dangerous dilettante).

Conclusions

Those who currently provide health care ethics consultation in Canada are a very heterogeneous group. There is a broad range of formal training and of time commitment to clinical ethics work, a broad range of expertise and wide variation in practices. This heterogeneity is probably much greater than that found among those who write and comment on ethics consultation. The latter authorities are more likely to be full-time practitioners who have their roots in a normative discipline. In general, ethics consultants seem to be content with a high degree of heterogeneity. There is at least no major thrust for change. Interestingly, a large proportion of the respondents to the questionnaire defined "bona fide" ethics consultants as having training and job definitions that few respondent themselves possessed. This suggests a perceived difference between those who are to be considered "ethics consultants" and those "who provide ethics consultation" by reason of their informal learning and experience and the need for some kind of ethics consultation in the institutions with which they are associated.

There is little information as to the views of the consumer on the issue. We do not know whether clinicians requesting ethics expertise would prefer the physician-ethicist model or would prefer the full-time expert from one of the normative disciplines. Similarly there is no evidence as to the relative effectiveness of these different models.

Physicians who are active in ethics consultation tend to be established practitioners. However, it has only recently been accepted that the potential exists for direct entry of clinical ethics consultants from early training. The numbers of such individuals, therefore, may as yet be small and this particular type of practitioner may not be well reflected by the survey. It seems likely that ethics consultation is in a transition period, during which we are seeing the emergence of a group of younger practitioners from both the normative and health-related disciplines and a relative decline of physician dominance. This would certainly mirror the emergence of interdisciplinary team decision making and management, which is an increasingly frequent form of practice in clinical medicine itself.

Educational backgrounds are regarded by those who practice consultation as important. The high proportion of consultants with multiple educational and professional backgrounds and advanced educational qualifications suggests that many ethics consultants enter practice after substantial amounts of education, professional, and general life experience. It is interesting to speculate, therefore, whether the new generation of somewhat younger individuals might respond differently to requests for consultation.

The emergence of the specifically trained younger individual can be compared to the early stages of the emergence of subspecialities in medicine. However, in contrast to the medical model where subspecialists tended to emerge from a similar or consistent background (for example, the subspecialty gastroenterologists having started training as a general intern or medicine specialist), in ethics consulting, we see a wide variety of initial expertise. As a result, the process of developing accreditation or certification is more difficult and uncertain. Given the absence of evidence that the specifically trained individual is either more effective or preferable to consumers, given the absence of agreement on the common pathway for training, and given the diversity of feeling expressed by current practitioners, it seems a reasonable conclusion to support the overall preference of the Network for identifying an ideal Profile for the end product, that may be only partially achievable, rather than defining a process for accreditation or certification of training.

Notes and References

[1]A number of studies have examined the qualities and training of ethics consultants, but not from an empirical perspective. *See,* for example, Terrence F. Ackerman, "The Role of an Ethicist in Health Care," *Health Care Ethics: A Guide for Decision-Makers,* eds. Gary R. Anderson and Valerie A. Glesnes-Anderson (Rockville, MD: Aspen, 1987) 308–320; Françoise Baylis, "Moral Experts and Moral Expertise: Wherein Lies the Difference?" *Clinical Ethics: Theory and Practice,* eds. Barry Hoffmaster, Benjamin Freedman, and Gwen Fraser (Clifton, NJ: Humana, 1989) 89–99; Peter A. Singer, Edmund D. Pellegrino, and Mark Siegler, "Ethics Committees and Consultants," *Journal of Clinical Ethics* 1 (1990): 263–267; John La Puma and David L. Schiedermayer, "Ethics Consultation: Skills, Roles, and Training," *Annals of Internal Medicine* 114.2 (1991): 155–160; David C. Thomasma, "Why Philosophers Should Offer Ethics Consultations," *Theoretical Medicine* 12 (1991): 129–140; John La Puma and E. Rush Priest, "Medical Staff Privileges for Ethics Consultants: An Institutional Model," *Quality Review Bulletin* 18. 1 (1992): 17–20.

[2]Society for Bioethics Consultation, 6th Annual Meeting, Chicago, Sep. 1992.

[3]Joyce Bermel, "Ethics Consultants: A Self-Portrait of Decision Makers," *Hastings Center Report* 15. 6 (1985): 2.

[4]Donnie J. Self and Joy D. Skeel, "Professional Liability (Malpractice) Coverage of Humanist Scholars Functioning as Clinical Medical Ethicists," *Journal of Medical Humanities and Bioethics* 9 (1988): 101–110. [After we submitted this chapter, a further report of the Self and Skeel study appeared: Joy D. Skeel, Donnie J. Self, and Roland T. Skeel, "A Description of Humanist Scholars Functioning as Ethicists in the Clinical Setting," *Cambridge Quarterly of Health Care Ethics* 2 (1993): 485–494.]

[5]Michael D. Coughlin and John L. Watts, "A Descriptive Study of the Health Care Ethics Consultant in Canada," *HEC Forum* 5 (1993): 144–164. The authors express their appreciation to the publisher of *HEC Forum* for permission to use material from that chapter in this chapter.

[6]Coughlin and Watts. The slight discrepancies in the data presented here and in the preliminary analysis are due to more careful screening for duplicate responses and to a different method for assigning "primary discipline" (i.e., "feeder" discipline) to the respondents.

[7]Peter Allatt, "The Ethics Committee Network of South Central Ontario (Canada)," *HEC Forum* 3 (1993): 212–216.

[8]Kindly provided by Professor Marie-Hélène Parizeau, coordinator of the GREM (Groupe de Recherche en Éthique Médicale). The study is published as Les Comités d'Éthique au Québec: Guide des Resources en Centres Hospitaliers (Montréal: Government of Québec, 1991).

[9]In the study by Self and Skeel, 15% of those providing ethics consultation were women.

[10]La Puma and Schiedermayer.

[11]It is important to note that until extremely recently, Canadian universities did not give doctorates in nursing.

[12]For a similar discussion, *see* Bermel.

APPENDIX

Strategic Research Network in Applied Ethics

"Ethics Consultation in Health Care"

Questionnaire on Health Care Ethics Consultation

1. Do you have, or have you had in the last five years, a role in providing ethics consultation in health care? (For the purposes of this questionnaire, health care ethics consultation includes consultation on ethical issues in clinical cases or in clinical research, ethics consultation to ethics committees, research ethics boards, and policy formulation committees in health care institutions.)

   ___ No There is no need to continue. Please return this questionnaire in the enclosed envelope.

   ___ Yes Please continue with the rest of the questionnaire.

2. Are you willing to provide your name for a list of health care ethics consultants in Canada (as defined above) that could be used to facilitate networking?

   ___ Yes ___ No

Your name will not be correlated with the data collected here, but we think it would be useful for networking to make available a list of those involved in health care ethics consultation.

If you are agreeable to having your name put on a list of those who see themselves as providing clinical ethics consultation, please fill in the appropriate sheet at the end of this questionnaire and put in the separate envelope marked List of Consultants.

I. Demographic Information

3. In what region of Canada do you work?

   ___ Atlantic region (Nfld., N.B., N.S., P.E.I.)

   ___ Eastern region (Que.)

   ___ Central region (Ont.)

   ___ Western region (Man., Sask., Alta., B.C.)

   ___ Territories

4. Age: ___ 20–29 ___ 30–39 ___ 40–49 ___ 50–59 ___ 60+

5. Sex: ___ Male ___ Female

Health care ethics consultation may mean different things to different people. One of the goals of this survey is to find out how people use the term. Some of the questions may not be applicable to your situation. Please answer those that are.

If you have done ethics consultation, but are no longer doing it, please have your answers reflect your position when providing consultation. If you have changed positions in the past five years and still provide consultation, please answer according to your current position. Please check as many answers as apply in each question.

II. Background Information

6. What is your professional/academic training?

___ Law degree(s):_________ ; year: ______

___ Medicine degree(s):_________ ; year: ______

___ Nursing degree(s):_________ ; year: ______

___ Other health profession

Specify:_________ ; degree(s):_________ ; year: ______

___ Philosophy degree(s):_________ ; year: ______

___ Science degree(s):_________ ; year: ______

___ Theology degree(s):_________ ; year: ______

___ Other

Specify:_________ ; degree(s):_________ ; year: ______

(or certification)

7. Do you have a degree with bioethics specialization?

___ No Go to #8.

___ Yes Answer below.

___ B.A. ___ M.A. ___ Ph.D. ___ Other.

Specify:___________

From what faculty was your degree?:

Did the degree include a clinical practicum?

___Yes ___No

Please go to question #10.

8. If you do not have a degree in bioethics, have you had a specific period of training or education in bioethics (e.g., fellowship in bioethics, internship, courses, and so forth in bioethics)?

___ No Go to #9.

___ Yes Indicate components below.

	Total length of time					Academic	Practical		
	wk	wk	mo	mo	yr	component	component	Full-	Part-
Faculty	1	2–4	1–6	7–12	1+	(Y or N)	(Y or N)	time	time
__ Health Sci.	__	__	__	__	__	__	__	__	__
__ Medicine	__	__	__	__	__	__	__	__	__
__ Nursing	__	__	__	__	__	__	__	__	__
__ Philosophy	__	__	__	__	__	__	__	__	__
__ Religion	__	__	__	__	__	__	__	__	__
__ Theology	__	__	__	__	__	__	__	__	__
__ Institute	__	__	__	__	__	__	__	__	__
__ Other	__	__	__	__	__	__	__	__	__
Specify:									

Please briefly describe this education:

9. If you have not had a specific period of training or education in bioethics, or if your training has been self-directed (e.g., seminars, membership on ethics committee, and so on), please describe the nature of your experience or education:

III. Position Providing Health Care Ethics Consultation

___ Current

___ Not current, but held within the past five years

10. Whom do you regard as your primary employer?

    ___ University

    ___ Hospital

    ___ Other health care facility

    Specify:

    ___ Bioethics center

    ___ Other organization

    Specify:

    ___ Self-employed

11. Do you have appointments in institutions other than your primary employer?

    ___ No

    ___ Yes Specify:

12. Do you have a title that refers directly to your clinical ethics work?

    ___ No

    ___ Yes What title do you use?

    ___ Ethics consultant

    ___ Ethicist

    ___ Clinical ethicist

    ___ Bioethicist

    ___ Ethics advisor

    ___ Other. Specify:

For the purposes of this questionnaire, the generic term ethics consultant will be used throughout.

13. What percentage of your work time (during a year) is spent in clinical ethics work?

    ___ %

14. What percentage of your clinical ethics work time is spent in:

    ___ Clinical case consultation (including related reports)

    ___ Ethics committee work

    ___ Clinical research consultation

    ___ Research ethics board

    ___ Policy formulation

    ___ Ethics education

    ___ Administration

    ___ Research and writing on bioethics other than above

    ___ Public speaking

    ___ Other. Specify:

15. Is the provision of health care ethics consultation part of your job description?

    ___ Yes ___ No

16. To what kind of institution(s)/organization(s) do you provide health care ethics consultation?

    ___ Teaching (university associated) hospital

    ___ Nonteaching hospital

___ Long-term care facility

___ University

___ Other organization. Specify:

17. Are you paid (or is your employer paid) by any of these institutions to do health care ethics consultation? Please circle the appropriate answer.

	Never <– – Sometimes – –> Always				
__ Teaching hospital	1	2	3	4	5
__ Nonteaching hospital	1	2	3	4	5
__ Long-term care facility	1	2	3	4	5
__ University	1	2	3	4	5
__ Other organization	1	2	3	4	5

18. In what capacity do you provide health care ethics consultation services?

___ As part of duties as a full-time ethics consultant

___ As part of duties as a part-time ethics consultant

___ As a member of an ethics committee

___ As a member of a research ethics board

___ As a function of another clinical or hospital position

Specify:

___ As a function of another academic position

Specify:

19. At whose request, in practice, do you provide health care ethics consultation?

___ Administration

___ Ethics committee

___ Families

___ Media

___ Nurses

___ Other health professionals

___ Patients

___ Physicians

___ Research ethics board

___ Social work

___ Other. Specify:

20. What kinds of health care ethics consultation do you provide?

___ Clinical case consultation

___ Clinical research consultation

___ Consultation to ethics committees

___ Consultation to research ethics boards

___ Policy formulation on ethical issues

___ Consultation to other ethics consultants

___ Other. Specify:

IV. Clinical Case Consultation

21. Do you distinguish formal from informal clinical case consultation?

___ No

___ Yes How?:

22. Approximately how many times per month do you act as an ethics consultant for a clinical case or provide an ethics discussion service?

    ___ times per month

23. Do you act as an ethics consultant on your own or as a member of a team?

    ___ Solo ___ As member of team ___ Both

24. What are the common ethical issues associated with clinical cases for which you have provided consultation? **Please mark the three (3) <u>most</u> common issues with an "x" and others as applicable with a "√".**

    ___ Truth-telling

    ___ Confidentiality

    ___ Informed consent

    ___ Surrogate decision making

    ___ Competency evaluation

    ___ Refusal of treatment

    ___ Request for futile treatment

    ___ Living wills/advance directives family conflicts

    ___ Child or elder abuse

    ___ Interprofessional conflicts

    ___ Decisions around withdrawal of treatment

    ___ Decisions concerning appropriate care for dying patients

    ___ Nutrition and hydration

___ Euthanasia

___ Use of restraints

___ Decisions regarding appropriate psychiatric treatment

___ Discharge planning for patients at risk

___ Reproductive issues

___ Termination of pregnancy

___ Organ donation

___ Clinical research related issues

___ HIV-related issues

___ Triage within and across institutions

___ Allocation of resources

___ Animal experimentation

___ Others. Specify:

25. What are you **asked for** in a clinical case consultation? Please circle the appropriate answer.

	Never <– – Sometimes – –> Always				
Information	1	2	3	4	5
Clarification	1	2	3	4	5
Analysis	1	2	3	4	5
Advice	1	2	3	4	5
Recommendations	1	2	3	4	5
Answers to questions	1	2	3	4	5
Make decisions	1	2	3	4	5
Confirmation of decisions	1	2	3	4	5
Comfort, support	1	2	3	4	5
Mediation/arbitration	1	2	3	4	5

Identify unrecognized issues	1	2	3	4	5
Legal status of issue	1	2	3	4	5
Legal opinion on issue	1	2	3	4	5
Creative alternatives	1	2	3	4	5
Other. Specify:					
______________________	1	2	3	4	5
______________________	1	2	3	4	5
______________________	1	2	3	4	5

26. What do you **provide** in a clinical case consultation? Please circle the appropriate answer.

	Never <--		Sometimes		--> Always
Information	1	2	3	4	5
Clarification	1	2	3	4	5
Analysis	1	2	3	4	5
Advice	1	2	3	4	5
Recommendations	1	2	3	4	5
Answers to questions	1	2	3	4	5
Make decisions	1	2	3	4	5
Confirmation of decisions	1	2	3	4	5
Comfort, support	1	2	3	4	5
Mediation/arbitration	1	2	3	4	5
Identify unrecognized issues	1	2	3	4	5
Legal status of issue	1	2	3	4	5
Legal opinion on issue	1	2	3	4	5
Creative alternatives	1	2	3	4	5

Other. Specify:

_______________ 1 2 3 4 5

_______________ 1 2 3 4 5

_______________ 1 2 3 4 5

27. Do you prepare written summaries of clinical ethics case consultations? Please circle the appropriate answer.

Never <– – Sometimes – –> Always

1 2 3 4 5

28. Are clinical ethics case consultations recorded in the patient's chart? Please circle the appropriate answer.

Never <– – Sometimes – –> Always

1 2 3 4 5

29. If consults are recorded in the patient's chart, how is this done? Check as many as apply.

___ Note written by the ethics consultant

___ Note written by attending physician

___ Other. Specify:

30. Is the consultation service evaluated (e.g., "Quality Assurance")?

___ Regularly ___ Occasionally ___ Rarely ___ Never

If you do not know answer or if question is not applicable, please go to #32.

31. If so, check as many as apply.

How?

___ Questionnaire or survey

___ Number of requests for consultation

___ Verbal feedback

___ Letters

___ Specific evaluation tool

___ Audit of chart notes

___ Other. Specify:

By whom?

___ By self

___ By employer or delegate

___ By peers

___ By ethics committee

___ By staff requesting consult

___ Other. Specify:

V. Research and Dissemination

32. Do you have protected time for research in health care ethics?

    ___ Yes No

33. Do you have an opportunity to do research in health care ethics?

    ___ Yes ___ No

34. In the past five years, have you received grants as principal investigator to do ethics research?

    ___ No

    ___ Yes How many? ______

    At a total value of ? $___________

35. Have you given talks presenting your own research in the past two years?

    ___ No

    ___ Yes How often?

    ___ per year

36. Have you published articles on health care ethics topics?

    ___ No

    ___ Yes How many in the past five years?

    ___ peer reviewed

    ___ not peer reviewed

37. Have you published books on health care ethics topics?

    ___ No

    ___ Yes How many?

    ___ authored

    ___ edited

VI. Other Areas

38. What do you consider your single most important function as an ethics consultant?

39. How many persons who function as "bona fide" ethics consultants in health care would you estimate there to be in Canada?

    ___ persons

Please define briefly how you understand "bona fide" ethics consultant:

40. Do you think there is a need for certification of individuals offering health care ethics consultation services (e.g., licensing exam, accreditation of programs, evaluation of training)?

___ Yes ___ No

Comments:

41. General comments:

Finally, if you have any personal or institutional documents or guidelines for ethics consultation that you follow, we would appreciate it if you could send us a copy.

Index